Understanding Maths

Basic Mathematics Explained

Aber Education
Adult Skills Series: Basic Skills in an Adult Context
Reading Comprehension 1
Reading Comprehension 2
Reading Comprehension 3
Cloze
Cloze 2: Cars and Transport
Cloze 3
Guided Reading and Writing
Punctuation 1
Punctuation 2
Punctuation 3
Spelling 1
Spelling 2
Grammar 1
Grammar 2
Family Life
Writing Dictionary
Applications and Forms

Money Management
Understanding the Numbers
    The First Steps in Managing Your Money
Back to the Black
    How to get out of debt and stay out of debt

Mathematics
Algebra: Basic algebra Explained

Morals and Values Education for Children
Aussie Tales Series

Life Skills Series
Survival Teen Island
Family Relationships
Bullying and Conflict
Self-Esteem: a manual for mentors
Self-Esteem and Values
Enhancing Self-esteem in adolescents
Hey Thompson: Developing self-esteem and resilience in secondary school
    students
Grief, Illness and other issues
The Eat Well Stay Slim Budget Cookbook

Many other titles in preparation

Aber Education
Changing Lives

Understanding Maths

Basic Mathematics Explained

Fourth edition

Dr Graham Lawler

Aber Publishing

978-1-84285-265-1 Understanding maths 4th ed

Website@ http://www.aber-publishing.co.uk

Note: The contents of this book are offered for the purposes of general guidance only
and no liability can be accepted for any loss or expense incurred as a result of relying in
particular circumstances on statements made in this book. Readers are advised to check
the current position with the appropriate authorities before entering into personal
arrangements.

Case studies in this book are entirely fictional and any resemblance to real persons or
organisations is entirely coincidental.

The moral rights of the author have been asserted.

Typeset by PDQ Typesetting, Newcastle-under-Lyme.
Printed and bound in Europe

Contents

Foreword
Welcome to the Joy of Maths!

Sir Arthur C Clarke: author of *2001 A Space Odyssey* and the first person to provide the vision for telecommunications

Mathematics is one of the most valuable inventions – or discoveries – of humankind. It can also have an aesthetic appeal equal to that of any creation in art or music. Perhaps this is what inspired the poetess to proclaim: 'Euclid alone hath looked at beauty bare'.

Much of mathematics, as of ordinary life, involves distinguishing between the possible and the impossible. And sometimes that decision is extraordinarily difficult to make, even in what appears to be very simple situations.

Let me give you a little problem that is both simple and apparently easy. It's the classic puzzle of the three utility companies, who want to provide service to three customers. The problem is, neither the electric, telephone or cable-television company will permit anyone to lay a line that crosses their own.

A few minutes of doodling will quickly demonstrate that there's no difficulty in taking two of the services to all three homes, and another to any two of them but the final connection isn't so easy. I once encountered an old gentleman who had spent *years* trying to solve the puzzle, creating the cat's cradle of circuitry. Apparently it had never occurred to him that there is no solution, and it is not hard to demonstrate that the feat is impossible. (I'll leave it as an exercise for the student...)

At this point, you might wonder what practical purpose is served by solving maths puzzles. Well, as my old friend Martin Gardener would confirm, playing with numbers has an appeal of its own that is appreciated by tens of thousands of men and women across the planet. Some, like Martin, have spent a lifetime inventing, promoting and enjoying the intricate mind-play with numbers.

But not everyone will go along with the brilliant mathematician G H Hardy's famous toast: 'Here's to pure mathematics may it never be of any use to anybody!'

In his own lifetime, Hardy's toast had a spectacular refutation. Without the purest of mathematics, the outcome of World War II might have been very different.

What happened at Bletchley Park during the War years is no longer classified. We now know that is where a handful of young geniuses – including Alan Turing, of whom Churchill said, 'When I told you to leave no stone unturned, I didn't expect you to take me quite so literally' broke the Enigma cipher. The result was that the Allies were privy to the German High Command's most secret plans. The Germans were absolutely convinced that Enigma was unbreakable, but it was cracked by a combination of advanced mathematics and the world's first electronic computer. And so a handful of equations helped to change the course of history...

Even if we leave the waging of wars and pushing the frontiers of knowledge to the brainiest among us, a basic understanding and proficiency in mathematics can certainly help us in many, mundane ways. In a century that is going to be dominated by science and technology more than ever before, maths proficiency will become as indispensable as the ability to read and write. Those who cannot relate to percentages, fractions, probabilities, and statistical projections would find themselves quite unable to navigate the knowledge society or to survive in the knowledge economy.

Yet it is highly worrying that precisely these skills are lacking among many educated adults in both the UK and USA. Successive surveys have shown how these two countries are lagging far behind other nations in their ability to understand and handle numbers.

As Graham Lawler can confirm, it is never too late to acquire these skills and to start enjoying the joy of maths. For those who missed the opportunity of learning maths formally in school, for whatever reasons, this book offers a valuable second chance. I cannot agree more with Graham when he says, 'In the 21st century, you will earn what you learn; if you don't learn much, don't expect to earn much.'

Today, nobody is proud of being illiterate. It is high time that otherwise educated people stopped flaunting their innumeracy.

Sir Arthur C Clarke
Colombo, Sri Lanka
20 February 2006

The late Sir Arthur C Clarke was kind enough to write the foreword for the third edition of this book. We have kept his foreword in this fourth edition since his words are erudite and still as applicable now as they were in 2006. It is also our small way of honouring the contribution made to humanity by a great man.

Preface to the Fourth Edition

To be in a position to publish a new edition of *Understanding Maths* is a great honour and I want to thank the thousands of people who bought the first three editions and in particular those who took the time and trouble to write complimentary reviews on Amazon. Thousands of people are now being empowered and benefiting both personally and professionally and this is a source of great pride to me as an author and to the whole team at Aber Publishing.

As I said in the first edition, believe in yourself, you do have talent. Work through this book and you will be able to use the techniques explained here and be able to apply them in your own work. When you improve your numeracy, you have something extra to offer your employer, you have more confidence and more self-esteem and you are on a par with every other numerate person, which puts you on an upward spiral of achievement.

All workers in the future will need to be numerate since if they are not, they may well be made redundant and this is no idle threat. At the start of the 21st century we began to see supposedly skilled people, particularly those in middle management, being laid off. Whilst this is a calamity for those people concerned, especially when they are in their middle years, it makes sense for the companies and employers concerned.

Expectations of employees have changed. For instance, even juniors in an organisation are expected to set up a spreadsheet, understand performance targets and percentage increases or decreases. More and more companies are now expecting operatives on the factory floor to do this type of work so it is vitally important. If a company or employer has bought your knowledge why should they keep on employing you? The only reason to keep employing you is because you have something new to offer. Where do you get that new item? You have to learn it. It does not mean that everyone has to go to university. But it does mean that when you are choosing a career, you need to think carefully. The one growth area is in 'problem solving'. People who can solve other people's problems and produce a high quality job are the only group who have seen a real increase in their income in the last 25 years. But they are a particular type of problem solver – people who use analysis, charts, diagrams and symbols to solve problems. People like engineers, human resource/personnel experts; these are the type of jobs that US expert Robert Reich calls 'symbolic analysts'.

In other words they reduce the problem to symbols, they solve it and then apply it back to the real world. This ability to reduce the problem to symbols is at the heart of what mathematics is all about. Studying mathematics at any level helps you to think strategically, it helps you to analyse, to develop solutions and to follow them through.

Once again I am indebted to my wife and partner Judith Lawler for putting up with the domestic disruption that writing a book like this causes and to the late Sir Arthur C Clarke, who took time, whilst literally not having much of it left, to support this book. I hope that the fact that such an eminent and famous science writer shows his support will help persuade many people that this really is important; by improving your knowledge by 10 per cent, you can stand out from the crowd and it really is worth the effort.

A book is a team effort and although it is my name on the cover I want to ensure the team of Jude, Pete and Marian are all aware of the debt of thanks I have to them all.

We have unashamedly stuck to the same successful formula used in the first three editions and have broken down each chapter into bite-size chunks. This makes the material easier to learn and memorise. Treat this book as a friend, something to return to when you are in need of a boost of knowledge and I hope you enjoy what is a truly fascinating subject.

Graham Lawler

Preface to the First Edition

This book was written in response to a need expressed to me by a number of my adult students. For whatever reasons, a number of my students felt they had missed out during their schooling and had gaps in their knowledge. Whilst they did not express a desire to repeat school exam mathematics, this lack of knowledge was having an effect on their studies. This fact was coupled with the growing realisation that there is no longer such a concept as a job for life. All forward-thinking people are in agreement that the achievers of the 21st century will be those people who keep their skills up to date. In real terms this will give them marketable skills from which they can earn a living. This is what the phrase 'lifelong learning' means in everyday life. It means that each of us has a personal responsibility to ourselves and our friends and family.

Mathematics has an identity crisis in Western society. People who are illiterate hide the fact that reading and writing is a problem for them, yet paradoxically many of us wear our poor understanding of mathematics as a badge of pride. How often have you heard people proudly proclaim that they cannot do mathematics?

There is a situation far worse than this – it is the fact that many adults, including teachers, say the same thing to children. The subtext of such a statement is that it is possible to avoid mathematics, and I do not believe that this is a desirable sustainable state.

All of the techniques and suggestions in this book are tried and tested with my students, all of whom are adults. If you are in a similar position, this book can help. Treat it as a friend, someone you can turn to when you need help. But most of all believe in yourself – you are capable of learning, and learning mathematics will open many new doors and opportunities in your life. When you are stuck, as no doubt you will be, follow the advice of one professor who said to me:

'enjoy the state of being stuck, revel in it, because it says one thing, you are human.'

The material in these chapters is there to help you, to get you unstuck. It is broken down in manageable chunks to make sure that you understand the underlying ideas behind the mathematics. In twenty years of teaching, I have come to believe one thing: everyone is capable of learning mathematics, at their level. When the mathematics is clearly explained, I have never found a student who has not enjoyed the experience of mastering the concepts behind the subject, and I hope this will be true for you.

I would like to put on record my thanks to my wife Judith Lawler for having the patience and stamina to put up with the domestic disruption that writing a book like this causes.

I hope that this book helps you and that lecturers and teachers will recommend it as a help to their students.

Graham Lawler

Place Value

One-minute overview — Place value is a term that refers to the value a figure has, when in different places in a number. For example, in 193 the place value of the 9 is 90. In other words it is 9 in the tens column. It is worth 9 lots of 10, or 9 × 10, and so makes 90. But if the 9 were in the units column, the place value would only be 9 units, or 9 × 1, making 9. In this chapter you will learn:

▶ quick methods for multiplying and dividing by 10, 100, and 1000

▶ how to calculate with directed numbers

▶ quick methods to work out calculations mentally.

Where numbers come from

Using Arabic numbers

The numbers we use in everyday life are referred to as Arabic numbers. This is because these numbers were brought to Europe by Arabic tradesmen. In fact, the numbers we use were developed in Asia, principally by Hindu mathematicians. Before Arabic numbers became so well used, Europe depended on Roman numerals. These are still in use today. If you look at the end of many television programmes you will see Roman numerals are used to give the date, for example MCMLXVIII. Imagine the problem of multiplying MCMLXVIII by MCMXIV.

The use of Arabic numerals made life a lot easier for tradespeople in particular, because for the first time there was a symbol for zero. The use of the 'nought' is important. It is a symbol that is both a place holder and one that stands for a value in its own right. For instance, in a number like 109, the central zero tells us that there is no tens value in this number.

Making mental calculations

In this chapter we will look at ways of calculating mentally. Even with the development of the calculator we still need to be able to calculate mentally, in order to be able to validate the accuracy of our answer. Suppose you are working out 15 × 4. You could split it into 10 × 4 and 5 × 4 and then add the answers. Since 10 × 4 is 40, this should give you an idea of the size of the answer that you are expecting. If your final answer is, say 6, then this should send out warning signals that you have made an error.

Quick methods in calculating are essential in the workplace and certainly help people on the promotion ladder. Time spent learning these quick methods and practising them now will pay you dividends later. People who can calculate quickly can do more in a given time than colleagues who cannot quickly solve a problem. In career development terms, people who show that they are not calculator dependent stand out very clearly in the workplace from others. There are times when resorting to a calculator may not instil confidence in your client.

Case study

Shaun is 53 and sells agricultural machinery for a multinational company to farmers and agricultural dealers. Unfortunately, Shaun is not quick at working out percentages. When he has to give a 10% discount he can become uncertain. His way around this is to carry a chart in his pocket that he reads off every time he needs a figure. The problem with this is that he has to constantly update the chart when prices change, and sometimes finds himself having to do this twice a month. Each time he has to do this, it takes half a day, this is not a good use of his time.

What is place value?

Place value means the value that a figure stands for when either standing alone, or when part of a larger number. When you were at school, as a young child, the teacher no doubt asked you to work out questions involving hundreds, tens and units. Repetitive use of these figures was meant to give you an understanding of the place value of hundreds, tens and units.

For example, what is the value of the underlined figure in 3<u>7</u>4 ?

D.P. (decimal point)

H	T	U	t'ths	h'ths	th'ths
3	7	4			

Fig. 1

In this example, the 7 must mean 7 tens, because it is in the tens column. Since 7 × 10 makes 70, this 7 stands for seventy. There is a 3 in the hundreds column and a 4 in the units column. The number is 374 and reads as three hundred and seventy four.

In the same way, we can give values to columns after the decimal point. Look at this example:

D.P. (decimal point)

H	T	U	t'ths	h'ths	th'ths
		0.	3	6	5

Fig. 2

In this case, we are dealing with a number smaller than 1. It is a decimal fraction and we read it as 'nought point three six five'. You will find that many people will read this as 'nought point three hundred and sixty five'. This is wrong, and a practice you should avoid.

Understanding column values

In the example for 0.365, we have placed the decimal point in a column with the zero. This is to stress that the decimal point does *not* move. It is the figures that move around the decimal point.

The first column to the right of the decimal point is the tenths column, so the value of the 3 in this column is $\frac{3}{10}$ — three tenths.

The second column to the right of the decimal point is the hundredths column, so the value of the 6 in this column is $\frac{6}{100}$ — six hundredths.

The third column is the thousandths column, so the value of the 5 in this column is $\frac{5}{1000}$ — five thousandths. A value like $\frac{5}{1000}$ is very small. When you hear engineers or mechanics referring to a tolerance of 'five thou', this is what they mean, a very small number indeed.

So 0.365 is actually $\frac{365}{1000}$. We always take the last column as the indicator of the value of the fraction.

Mathematics and the car salesman

Edward James told us of a case he has uncovered, that occurred during the 1960s in North Wales. A man wanted to buy a car, at a time when a working man's take home pay would have been about £4 a week.

The man said to the salesman, 'before you say anything, I can only afford £80 a month at the most, and that will take all of my savings'.

The salesman simply said, 'Leave it to us'. They did some quick calculating and said 'We could do it at a stretch for £79 10 shillings (£79.50)'. The customer smiled and signed on the dotted line.

Had he said nothing, his monthly payments would have been in the region of £16. As Edward James put it, '*Who says you don't need mathematics!*'

Multiplying by 10

For many adults, multiplying by 10 can prove surprisingly difficult. There are supposed to be shortcuts like 'add a nought' — but it is important to understand the mathematics that is going on here.

Take the example, 13 × 10. The answer is 130, but let's look at what is going on in the columns.

When 13 is multiplied by 10, all of the numbers move one place to the left. The last column (which would otherwise be left empty) is filled with a zero. **The main point is that, when we multiply by 10, all of the figures move one place to the left.**

Multiplying decimals

When multiplying decimals by 10, remember that once again they move one place to the left.

D.P.

H	T	U	t'ths	h'ths	th'ths
		0.	6	8	9

Fig. 3

D.P.

		0.	6	8	9
		6.	8	9	

Fig. 4

In this example, what is 0.689 × 10?

Look at the columns: we know that each of the numbers moves one place to the left, so in your mind, move them one place to the left. What is the answer?

As before, every number moves one place to the left, so 0.689 × 10 becomes 6.89. Notice that you don't have to put a zero after the 9, because this does not affect the value of the number 6.89. In other words 6.89 is exactly the same as 6.890, or 6.8900, or 6.89000 and so on.

Multiplying by 100

What happens when you multiply 1 by 100? Make a mental picture of what happens in the columns. This time, the numbers move two places to the left. This is an example in mathematics where you need to think about what you know and apply this knowledge to a new situation.

Here, we see that the figure 1 moves two places to the left. This is true for any numbers in the columns. For instance, look at 0.234 × 100. Try and imagine the columns in your mind. All of the numbers move two places to the left, so the answer to 0.234 × 100 is 23.4. Check this on your calculator and make sure you agree! Visualise this for another calculation, say 12.456 × 100. Once again, visualise the columns and see the numbers moving two places to the left. The answer is 1245.6.

The beauty of this method is the speed. It enables you to work very quickly and accurately. This is what will make you stand out in comparison to others. You will be able to do more than your colleagues in the same time. Regardless of what job you do, or what course you are studying, this will be a real benefit.

Steve goes on an engineering course

Steve is 44 and lives in the West Country. He is studying refrigeration engineering and wants to start his own business. He said, '*They probably did teach me these methods in school, but that was 30 years ago and I didn't listen. On my course, I had to show the others how to multiply like this. It was great being able to do it so quickly.*'

Multiplying by 1000

Once again, think about what you already know and can remember from your school days.

▶ What is the effect on numbers when you multiply by 1000?

▶ What happens when you want to multiply 3 × 1000?

▶ What happens to the figures when you multiply by 1000?

You should see that whatever number is multiplied by 1000, then its figures move three places to the left.

So, for example, with 230 × 1000. Visualise this in the columns: in your mind's eye move the 2 three places left and then the 3 and then the 0 and then fill in any gaps with zeros. So 230 × 1000 must be 230 000.

Writing numbers

There has been a subtle change in the way numbers are written. If you were in school in the 1970s or earlier then you will have used commas to separate thousands. For example, 1000 was written as 1,000.

This is no longer the case. Numbers up to 9999 are generally written all together as in 9999, without spaces or commas. But the number ten thousand is written as 10 000 — notice the gap between the 10 and the three zeros that follow.

The comma is no longer used, because in the rest of Europe a comma is often used as a decimal point. For example, in Europe 1,34 would read as 'one point three four'. Europeans also commonly use the point to indicate the process of multiplication (mathematically correct), so 1,34.5 means 'one point three four multiplied by five'.

So, the general rule when writing numbers over 10 000 is to batch the digits in threes. For example four hundred thousand is written as 400 000.

Dividing by 10, 100, 1000

We can use the columns to understand the effect of dividing by 10, 100, and 1000.

For example, look at the number 40. Obviously, from life experience you know that 40 ÷ 10 = 4, so what happens to the figures when you divide by 10? Look at where the 4 was when you wrote 40 and where the 4 is now in the answer. What do you notice? In the columns 4 has moved one place to the right. Try this with other numbers, until you are satisfied this is always true.

What about dividing numbers by 100? Try some numbers of your own in the columns. Again, what do you notice?

D.P.

H	T	U	t'ths	h'ths	th'ths
	4	0			
		4			

Fig. 5

▶ Does this always happen?

▶ What about numbers that are divided by 1000?

▶ Explain what happens to the numbers in the columns now.

Try experimenting with decimal numbers and check your answer on your calculator. Take a number – say 0.425 – and divide that by 1000. What happens? Explain, to another person, what is happening. Does the same thing always happen when you divide a number by 1000? Confirm your results with a calculator and write down the rules for dividing by 10, 100, and 1000.

If you have not already worked through the piece of work on the previous page, please do so now. It will really help develop your understanding of the mathematics.

Starting investigative work

The piece of work that you have just worked through is called a **mathematical investigation**. This style of learning is becoming more widespread in schools and is very effective in helping you to learn the mathematics. It may not have been in evidence when you were at school (especially if that was during the 1970s and early 1980s) yet it works extremely well.

This style of learning helps the learner to construct his or her own understanding. Traditionally it was thought that all the teacher had to do was to tell the student and then s/he would understand. We now know enough about learning to know that this is not the case. In short, the best way to learn maths is first of all to do some maths, then to reflect on what exactly you did, and so to understand the underlying principles.

Back to the investigation

Going back to the short investigation, you should have found that

▶ *dividing* by 10 always pushes the figures one column to the right.

▶ *multiplying* by 10 always pushes the figures one column to the left.

When you divided by 100 the numbers moved two columns to the right; when you divided by 1000, you should have found that the numbers moved three columns to the right.

These are the basic rules for dividing quickly by 10, 100, and 1000. Given practice and persistence you can master these skills very quickly.

Avoid trying to learn rules like, 'to multiply by 10, add a nought'. Quick fix rules like this often lead students to become confused. It is far better to invest the time and to understand how the columns work for multiplying and for dividing numbers.

Directed numbers

Directed numbers are **positive** and **negative** numbers. Numbers that are negative always have a − sign in front of them. Numbers that are positive may have a + sign in front of them, or there may be no sign. If there is no sign showing, the number is positive. This can be a real trauma for many children at school. If it was for you, during your school career, then simply recognise that you were a normal child then and no doubt you are a normal adult now.

Small children are taught that numbers start at 0 and go on and on for ever. This is because they are too young to grasp the idea of a negative quantity. Yet as adults, we deal with negative quantities every day. For instance, when you have to scrape ice off the car in winter, the temperature is a negative quantity, such as three degrees below zero. When you have an overdraft with the bank then your account is a negative quantity. For example if you have a fifty-pound overdraft, then you have an account of $-£50$.

Writing negative quantities

Dealing with negative quantities is complicated by the fact that we use the $-$ sign to indicate subtraction, and also to stand for the magnitude (size) of a quantity.

It is important to understand the difference between the two, but unfortunately some publishers have made it difficult over the years.

▶ Strictly speaking, the subtraction operation should be referred to as 'minus', for example $10 - 3$ (10 minus 3).

▶ On the other hand a negative quantity like -3 should be referred to as 'negative three'. The negative sign should be in superscript like this $^-$ and not as so often happens like this $-$.

So the question 'four minus negative three' should really be written as $4 - {}^-3$.

By the location of the $-$ signs on the page, it becomes quite easy to see which $-$ sign refers to the operation and which refers to the magnitude (size) of the quantity.

Adding quantities involving negative numbers

In order to understand this, think about what happens in everyday life. Imagine you have an overdraft of £300 and you put £500 into your account. What is the status of your account now?

In mathematical terms, this is written as:

$$^-£300 + £500$$

and will give you an answer of £200.

Using a calculator with negative numbers

On your calculator you will see there is a button marked either $+/-$ or $-/+$ When you want to enter a negative value in your calculator enter the positive value first, and then press this button. It will change the value to a negative for you. For instance, if you wanted to enter ⁻5, simply key in 5. Then press the $+/-$ or $-/+$ button on your machine (your machine will only have one of these buttons) and this will change the positive quantity into a negative quantity.

Adding negative numbers

When you have to add negative numbers, think of it as a ladder. Adding always takes you *up* the ladder.

Look at the ladder: instead of 0 being the start of all whole numbers, in fact it is in the middle. Then we have positive numbers that move upwards, in other words they get bigger. It also means that we have negative numbers that go down, or get smaller.

As you go down the ladder, the operation is **subtraction**. As you go up the ladder the operation is **addition**. So $4 - 6$ means that you are at ⁺4 and you go down 6. Look at the diagram: what is the answer?

Clearly the answer must be ⁻2, in other words, $4 - 6 = ⁻2$.

Fig. 6

What about adding a positive value to a negative number?

Again, imagine your own ladder or look at the ladder above and try to answer ⁻6 + 5. You must be at the sixth step below zero and are going up 5 places, so the answer must be ⁻1, in other words ⁻6 + 5 = ⁻1.

Try some examples of your own, to make it clear to yourself. Part of learning mathematics successfully is to convince yourself first and then convince someone else. It is worth discussing this with someone: it will help your understanding.

James Craig is a writer and mathematics teacher. He told us of a former student of his who, in the absence of anyone to discuss her mathematics with, spent two hours explaining the maths to the goldfish. The student felt that the experience helped her to understand the mathematics; the reaction of the goldfish is unknown.

Subtracting with negative numbers

In the same way as adding, we subtract from negative numbers by moving down the ladder. For instance, $^-6-5=$?

Imagine the ladder: you are at $^-6$ and then go down another 5, where are you now?

Think about this for a moment before you look at our answer.

You were at 6. You've gone down another 5 so you must be at 11. As before, try some questions of your own and convince yourself, and then try convincing someone else — or the goldfish!

Multiplying negative numbers

Rules

The mathematics behind some of this is actually quite complex and beyond the scope of this book, so you need to learn certain rules. (This contradicts what we said earlier about understanding the underlying mathematical principles, but please bear with us.)

In simple terms, when multiplying two numbers, if the signs are the same, the answer is a plus ($+$). If the signs are different, the answer is a minus ($-$).

So $^-4 \times {}^-3 = {}^+12$. Here the 4 is a negative quantity, and the 3 is a negative quantity, so the answer is a positive quantity.

What about $^+4 \times {}^-3$? Here, the signs are different. One is positive and one is negative, so the answer is negative: $^+4 \times {}^-3 = {}^-12$.

Rules summary

In easy terms:

$+ \times +$	$= +$
$- \times +$	$= -$
$+ \times -$	$= -$
$- \times -$	$= +$

More on adding and subtracting

Earlier we looked at problems like $^-3 + 4$. We said that we are three below zero and go up 4, so we must be at $^+1$.

Another way to write this would be as, $^-3 + {}^+4$. Here the two $+$ signs in the centre follow the same rules as above. In other words, the two signs are the same (in the middle, they are both + signs) and so the question is as first stated: ⁻3 + 4.

Students sometimes confuse this issue with the two signs. They sometimes will look at a problem like ⁻3 + 4 and see a negative in front of the 3, a positive in front of the four, and wrongly think that since these two signs are different, the answer must be negative.

Make sure that you are clear on the difference in your own mind. Earlier we were multiplying, so ⁻3 × ⁻4 = ⁺12.

BUT, when adding or subtracting you consider the two signs in the middle of the problem only. Here we have made them bold, and larger than normal, to stand out:

$$⁻3 + ⁺4 = ⁻3 + 4 = 1$$

Similarly ⁻3 + ⁻4 = ⁻3 − 4. The two signs in the middle are different, therefore they become − and ⁻3 − 4 = ⁻7. Think about it. You are at the third rung on the ladder below zero and you go down another four rungs, so you must now be at the seventh rung below zero, or ⁻7.

Negative numbers in real life

What if the temperature on a particular day was ⁻7°C and the following day it was four degrees lower? What was the temperature on the second day?

Here the mathematics is ⁻7 − ⁺4 and this is the same as ⁻7 − 4 = ⁻11. So the temperature on the second day was ⁻11°C (Brrr, very cold).

Dividing negative numbers

Some mathematics educators do not bother to teach this part of directed numbers, believing that students should use a calculator to do so. Yet many examination boards are now insisting on a non-calculator paper; the actual mathematics is quite straightforward here, so that, taken carefully, it should not trouble you.

Example
Let's consider:

$$⁻6 ÷ ⁻2$$

One of the easiest ways to think of this is in terms of a multiplication. Ask yourself: what do I need to multiply $^-2$ by, in order to generate an answer of $^-6$? Well, it must be $^+3$, because $^+3 \times {}^-2 = {}^-6$, so my answer is $^+3$.

Try this one:

$$^-20 \div {}^+5$$

Ask yourself, what do I need to multiply $^+5$ by to make $^-20$? The answer must be $^-4$ because $^+5 \times {}^-4 = {}^-20$.

Practice questions

1. $4 - 7 =$
2. $5 - 9 =$
3. $12 - 19 =$
4. $17 \times {}^-4 =$
5. $20 \times {}^-5 =$
6. $50 \times {}^-2 =$
7. $120 \div {}^-12 =$
8. $42 \div {}^-7 =$
9. $160 \times {}^-3 =$
10. Check the working of this calculation and see if it is correct. If it is incorrect, say where the error occurs and state the correct answer:

    $$^-6 \times {}^-9 = {}^-54$$

    Now add 7 and the answer is $^-47$. Add another 5 and the answer is $^-42$. Divide by 6 and the answer is $^-7$.

Mental maths

Do you remember the standing joke when you were at school, 'We're doing mental maths to see if you are mental'? Working something out mentally is important for speed. It is possible to learn different techniques that can help you to work quickly.

Jakow Trachtenburg's amazing techniques

Trachtenburg was a brilliant engineer who was incarcerated in Hitler's concentration camps. To save his sanity he developed amazing techniques to aid calculation. Trachtenburg was convinced that talent, in terms of mathematical ability, was constantly being lost. In the early years of elementary education, children developed negative attitudes to mathematics. He called this an emotional roadblock.

How true this still is, today. Trachtenburg's techniques are beyond the scope of this book but it is a fact that an 11-year-old child can successfully be taught to multiply 5623487958254123 by 11, and get the right answer in less than 30 seconds.

We are not expecting you to perform such amazing feats at the moment, but there are simple, very effective things you can do. The key point is, do not let yourself be held up by an emotional road-block. You know your name, you know your address, you know lots of other things that you once did not know, like the names of famous people, quotes from Shakespeare perhaps or the names of famous footballers. You know these because in the past you have learned this information. In the same way, with the help of this book, you can also learn the foundations of mathematics.

Splitting numbers

One good technique for multiplication is to split big numbers into smaller ones. For example, 15×4 can cause some problems for some people. Work out 10×4, then 5×4, and get the final answer by adding the two together:

$$10 \times 4 = 40 \qquad 5 \times 4 = 20 \qquad 40 + 20 = 60$$

So, $15 \times 4 = 60$. Try some questions of your own. Check them with a calculator to make sure that they work.

Tips for dealing with times tables

You may have memories of your primary school teacher shouting at you to learn your tables, perhaps feeling threatened but not knowing what to do. One thing that you can do as an adult is to learn certain answers in a times table, and then work out others from that answer. For instance, if you learn that $5 \times 7 = 35$, then it is not difficult to work out 8×7:

$$5 \times 7 = 35 \qquad 3 \times 7 = 21$$

So 8×7 must be $35 + 21 = 56$.

Nine times table

It is possible to work out any of the nine times table from your hands:

Fig. 7

Say for instance you wanted to work out 3×9. Drop the third finger on your left hand. The numbers in front of the dropped finger are the tens, and the numbers after the dropped finger are the units. Since you have two fingers up in front of the dropped finger, that is the two tens (i.e. 20). There are seven fingers after the dropped finger, so 3×9 must equal 27.

Try this for other nine times table calculations. It does always work.

Tutorial

Progress questions

1. Work out:
   (a) $235.36 \div 100$ (b) 156×10
2. Work out:
   (a) $^-4 \times {}^+5$ (b) $23 + {}^-8$

3. Work out:
   (a) $^-16 \div {}^+4$ (b) $^-9 - 9$

Seminar discussion

1. 'The argument over whether the calculator is a positive aid in helping people learn mathematics is being overshadowed by a

political debate on the style and type of education that dominates within the country.

Whenever new technology has been introduced, there have always been those who oppose it, those who feel it threatens the very fabric of our society and those who are just ignorant. The point is, the calculator relieves a lot of drudgery and empowers learners to become more confident and to learn more mathematics. But that is not an excuse for students to become dependent on the calculator for mundane calculations like 3 × 2.' Do you agree?

2. 'Mathematics is something that should be removed from the school curriculum and be available only to a small number of people who actively wish to study it. Most people only need to know how to add up numbers.' Do you agree? If so, why, and if not, why not?

Practical assignments

1. Set up a spreadsheet on a computer, to monitor the spending from your household. You will need to have two columns per month, one an estimate of expenditure and the other the actual expenditure.

2. Pick an aspect of your regular leisure expenditure, e.g. visits to the local public house. Estimate the frequency of your visits and an average cost per visit. Work this out for one year. Now divide this figure by 10 and consider opening an account to save $\frac{1}{10}$ of this budget.

Study and revision tips

1. Spend time and practise manipulating numbers. Investigate things for yourself, to build up your own confidence. For example, if an even number is multiplied by an even number, what sort of number is the answer? Does this always happen?

2. What about an odd number multiplied by an even, is this the same or different to an even number multiplied by an odd one?

3. Carefully read through the booklet that comes with your calculator and ensure you understand.

4. In your notes, show the working out and then in another colour, say red, write reminders to yourself about how you worked out the question.

5. At least three times weekly read through your notes and make sure you understand the work that you have done in your course.

6. Read around a subject. For instance, see what you can discover about the history of maths — it is fascinating.

2

Dealing with Fractions

One-minute overview – Fractions are a cause of anxiety to many people, because they did not develop their full understanding of fractions as a child. However, with a little bit of thought and some practice, it is possible to understand and manipulate fractions well. Many adults feel that it is no longer appropriate to learn how to work with fractions, because they have a calculator or computer. But as we mentioned in Chapter 1, it will pay you to be independent of the calculator, and think for yourself. There are certain jobs where you need to be able to work out fractions of quantities very quickly, so practice is very important. There are many different ways of calculating with fractions, but in this chapter we focus on the more straightforward ones. In this chapter, you will learn about:

► equivalence of fractions

► how to add, subtract, divide, and multiply fractions

► how to convert between improper fractions and mixed numbers.

Fractions often cause older people to snort with derision, and make statements of how things were always better in their day. This is fiction. Older people often hark back to a supposed golden age where things were always better. In reality there has never been such an age. This can cause problems for some adult learners, because it can destroy any confidence that they may have, so be aware of it, and make sure that they do not knock you off course.

What is a fraction?

A fraction means a part of a whole. Any whole that is broken into parts is said to be split into fractions. For example, $\frac{1}{2}$ is a fraction and means one whole split into two.

Showing fractions

Look at Figure 8. It shows a whole split into thirds, so clearly there must be three thirds in a whole. Mathematically this is written as $\frac{3}{3}$.

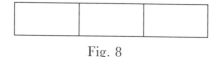

Fig. 8

This is true for any fraction, for example:

$$1 = \frac{2}{2} = \frac{3}{3} = \frac{4}{4} = \frac{5}{5} = \frac{6}{6} = \frac{100}{100} \quad \text{and so on}$$

Naming the parts of a fraction

The top of a fraction is called the **numerator**, and the bottom part is called the **denominator:**

$$\frac{numerator}{denominator}$$

It is also worth noting that in the UK the fraction $\frac{1}{4}$ is called 'one quarter'. This term is commonly used in everyday life, for example 'a quarter of a pound'. But in the USA this fraction is known as 'one fourth'.

What is the connection between the denominator and the size of a fraction?
It often bewilders some people why some fractions are larger than others.

For instance, how can $\frac{1}{2}$ be bigger than $\frac{1}{4}$, when 4 is a bigger number than 2?

Look at Figure 9. Imagine this diagram represents two cakes. The first cake has been cut into halves, whereas the second cake is cut into quarters or fourths.

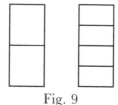

Fig. 9

You can easily see that the size of a slice of the cake cut into halves is bigger — in fact twice the size — than a slice of the cake that is cut into quarters. So, in simple terms, the larger the denominator of a fraction, the smaller the actual fraction. For example, $\frac{1}{6}$ is smaller than $\frac{1}{4}$ which in turn is smaller than $\frac{1}{2}$.

Think about it in terms of the slices of the cake. The more slices that the cake is split into, the smaller must be each slice.

What happens if I multiply the numerator and denominator by the same number?
Look at this fraction, $\frac{1}{2}$. Multiply the numerator and the denominator by 2. You get:

$$\frac{1 \times 2}{2 \times 2} = \frac{2}{4}$$

We have already seen how $\frac{1}{2}$ and $\frac{2}{4}$ are the same fraction, so the effect of multiplying both top and bottom by the same number is to leave the value of the fraction unchanged. This is a very useful technique in mathematics at a higher level, especially when dealing with fractions in algebra.

$$\times 2$$
$$\frac{1}{2} = \frac{2}{4}$$
$$\times 2$$

Fig. 10

Equivalent fractions
Again, using the cake analogy, we can cut the cake so that, say, two slices of one cake are the same as one slice of another cake.

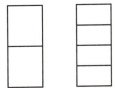

Fig. 11

This diagram shows the two cakes cut into halves and quarters. From the diagram it is easy to see that $\frac{1}{2}$ is the same as $\frac{2}{4}$. However, it is unrealistic to constantly have to draw diagrams like this to find equivalent fractions. There is an easier way, based on number patterns.

Families of fractions

Fractions that are **equivalent** can be said to belong to a 'family'. In other words, all fractions that belong to a 'family' are equal. Take for instance the family of halves:

$$\frac{1}{2}$$

Now multiply the numerator (top) and denominator (bottom) by 2, so that:

$$\frac{1}{2} = \frac{2}{4}$$

In this case, we multiplied the numerator and the denominator of the fraction by the same number, namely 2. Now try multiplying by 3, so that means:

$$\frac{1}{2} = \frac{3}{6}$$

and we can carry on:

$$\frac{1}{2} = \frac{2}{4} = \frac{3}{6} = \frac{4}{8}$$

and so on. Compare the numerator and the denominator of each of the fractions, what do you notice ? You should notice that the top of the fraction is half of the bottom of the fraction.

In other words, whenever a numerator is half of the denominator, the fraction is equal to a half.

'Family of thirds'

Similarly for the 'family' of thirds, start with $\frac{1}{3}$ and then multiply the numerator and the denominator by the same number, and then build up the family of fractions. So:

$$\frac{1}{3} = \frac{2}{6} = \frac{3}{9} = \frac{4}{12}$$

and so on; again, each of these fractions are equal. Notice the top of the fraction is a third of the bottom of the fraction, so whenever you have a fraction with a numerator that is a third of the denominator, then the fraction must be equal to one third.

Dealing with larger fractions

Larger fractions can have their equivalent fractions worked in the same way, for example:

$$\frac{2}{3}$$

Multiply the top of the fraction and the bottom by the same number, say 2, so that we have:

$$\frac{2}{3} = \frac{4}{6}$$

Try some for yourself. What other fractions can you find that are equal to $\frac{2}{3}$?

Study tip

On most new calculators you will find a button marked $a\frac{b}{c}$. This is a fraction button. To enter a fraction like $\frac{3}{4}$ you key in 3, then press the $a\frac{b}{c}$. button, and then key in 4. You will see a display that will show the 3 and then an L shape on its side and then the 4.

Try entering a fraction like $\frac{6}{8}$ and then press the = button. What happens? Try this for other fractions. You should find that the calculator cancels the fraction down for you. It is worth spending some time learning how your calculator works; it will repay you in your course.

How to add fractions

In this example we are going to add $\frac{1}{3}$ and $\frac{1}{2}$. These two fractions do not have the same denominator, so cannot be added yet. They have to be converted to have the same denominator.

This means that we need to find the family of fractions for $\frac{1}{3}$ and then for $\frac{1}{2}$.

Then, when we have the equivalence, we can add the fractions like this:

$$\tfrac{1}{3} = \tfrac{2}{6} = \tfrac{3}{9} = \tfrac{4}{12}$$

and

$$\tfrac{1}{2} = \tfrac{2}{4} = \tfrac{3}{6} = \tfrac{4}{8}$$

Now look along both family lines to find the lowest denominator that is in both families. Clearly, this is 6. In other words $\tfrac{1}{2}$ and $\tfrac{1}{3}$ can both be converted into fractions of sixths like this:

$$\tfrac{1}{3} + \tfrac{1}{2} = \tfrac{2}{6} + \tfrac{3}{6} = \tfrac{5}{6}$$

Here we have changed the fractions into sixths and then added them.

Incidentally, most adults cannot do this in their heads, so if you found it a little challenging, go back and read through the explanation again.

Let's try another example

$$\tfrac{1}{4} + \tfrac{1}{5}$$

Again, work out the family of fractions for each of these fractions

$$\tfrac{1}{4} = \tfrac{2}{8} = \tfrac{3}{12} = \tfrac{4}{16} = \tfrac{5}{20} = \tfrac{6}{24}$$

$$\tfrac{1}{5} = \tfrac{2}{10} = \tfrac{3}{15} = \tfrac{4}{20}$$

You can see that each fraction can be converted to a fraction over 20, so:

$$\tfrac{1}{4} + \tfrac{1}{5} = \tfrac{5}{20} + \tfrac{4}{20} = \tfrac{9}{20}$$

$$\tfrac{1}{4} + \tfrac{1}{5}$$

Warning

It is not simply a case of multiplying the denominators to find out what the fractions convert to.

For example suppose you needed to add $\tfrac{1}{4}$ and $\tfrac{1}{2}$. Here, if you multiplied the two denominators, you would convert these fractions into eighths, but 8 is not the **lowest common denominator**. Do you remember this term from school? It means the lowest denominator that can occur in both fractions.

For example, if you write out both sets of families of fractions for $\frac{1}{4}$ and for $\frac{1}{2}$ they are:

$$\frac{1}{2} = \frac{2}{4} = \frac{3}{6} = \frac{4}{8}$$
$$\frac{1}{4} = \frac{2}{8}$$

You can see in this list that 8 is a denominator that is common to both families of fractions but it is not the *lowest* common denominator.

In fact, the lowest common denominator is 4. In other words, 4 is the lowest denominator in the family of fractions of $\frac{1}{2}$ that is also in the family of fractions of $\frac{1}{4}$.

Improper fractions

An improper fraction is commonly called a **top heavy** fraction. This is an example:

$$\frac{3}{2}$$

But what does a fraction like this mean?
The fraction above actually means three halves, which is obviously the same as $1\frac{1}{2}$. Take a look at Figure 11. It shows a whole split into two halves and another half, so that you have three halves, or $\frac{3}{2}$.

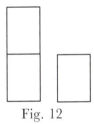

Fig. 12

The same is true for any improper fraction. You probably remember from school that one can be written as any number over itself, for example $\frac{3}{3} = 1$, $\frac{5}{5} = 1$ and so on. This means that whenever we are dealing with improper fractions, it is straightforward to understand exactly what the fraction represents.

For example, $\frac{5}{4}$ means 5 quarters. In diagrammatic form, it looks like this:

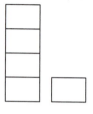

Fig. 13

What are mixed numbers?

Mixed numbers are whole numbers and fractions. For example $3\frac{1}{2}$ is a mixed number, and so is $113\frac{1}{4}$.

Changing improper fractions to mixed numbers

When you have to convert from an improper fraction to a mixed number, think of dividing the numerator by the denominator. Take $\frac{5}{4}$:

Step one $5 \div 4 = 1$ remainder 1

Step two   Since you are dividing by 4, that remainder 1 is actually $\frac{1}{4}$, so $\frac{5}{4} = 1\frac{1}{4}$.

Another example

Change $\frac{9}{5}$ to a mixed number:

Step one $9 \div 5 = 1$ remainder 4

Step two   Since you are actually dividing by 5, the remainder is actually $\frac{4}{5}$

So: $\frac{9}{5} = 1\frac{4}{5}$

Changing from mixed numbers to improper fractions

Think of a mixed number in terms of its diagram, for example think of $1\frac{1}{2}$ as:

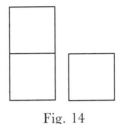

Fig. 14

Here you can see that $1\frac{1}{2}$ must be the same as $\frac{3}{2}$.

Let's try another but this time without drawing the diagrams:

$$5\frac{1}{4}$$

Now let's think through this logically, together. We know that 5 represents 5 whole ones. If each of those five is cut into quarters, then there must be 20 quarters in 5 (5 × 4). Now we need to add the other quarter (we are dealing with $5\frac{1}{4}$), so altogether $5\frac{1}{4}$ must equal 21 quarters. This is written as $\frac{21}{4}$.

How to subtract fractions

In mathematics there are often many techniques you can use to solve a particular problem or to answer a particular question.

Throughout this book we have tried to use the most straightforward ones, but they are not always the quickest. Please bear with this, because it is important for your learning to understand the *processes* that are occurring in the mathematics.

We have already discussed how to develop a family of fractions for a particular fraction, above. This is also a useful technique when we need to subtract fractions.

Take for example:

$$\frac{3}{4} - \frac{1}{3}$$

First of all, find the first few members of the family of fractions of $\frac{3}{4}$:

$$\frac{3}{4} = \frac{6}{8} = \frac{9}{12} = \frac{12}{16}$$

Now do the same for $\frac{1}{3}$:

$$\frac{1}{3} = \frac{2}{6} = \frac{3}{9} = \frac{4}{12} = \frac{5}{15}$$

Next, look down the row of denominators: find the lowest number that appears in both rows, in other words the lowest common denominator. In this case it is 12. So:

$$\frac{3}{4} = \frac{9}{12} \text{ and } \frac{1}{3} = \frac{4}{12}$$

$$\frac{3}{4} - \frac{1}{3} = \frac{9}{12} - \frac{4}{12} = \frac{5}{12}$$

Try it on your calculator to confirm the answer.

Adding and subtracting with mixed numbers

This is more straightforward than it first appears. Take the example:

$$10\frac{1}{2} - 3\frac{3}{8}$$

Step one: work out the whole numbers first, namely $10 - 3 = 7$

Step two: work out the fractions in the same way as we did earlier:

$$\frac{1}{2} - \frac{3}{8} =$$

family of fractions for $\frac{1}{2}$

$$\frac{1}{2} = \frac{2}{4} = \frac{3}{6} = \frac{4}{8}$$

Notice, since the family of fractions of $\frac{1}{2}$ contains a denominator of 8, there is no need to develop the family of fractions of $\frac{3}{8}$:

$$\frac{1}{2} - \frac{3}{8} = \frac{4}{8} - \frac{3}{8} = \frac{1}{8}$$

Step three: Don't forget to add the 7 that we worked out earlier. So the final answer is $7\frac{1}{8}$.

Let's try another example

$$12\tfrac{1}{4} - 4\tfrac{1}{3}$$

Step one: As before, work out the whole numbers first:

$$12 - 4 = 8$$

Step two: Work out the fractions:

$$\tfrac{1}{4} - \tfrac{1}{3}$$

The family of fractions of $\tfrac{1}{4}$ are: $\tfrac{1}{4} = \tfrac{2}{8} = \tfrac{3}{12} = \tfrac{4}{16}$

The family of fractions of $\tfrac{1}{3}$ are: $\tfrac{1}{3} = \tfrac{2}{6} = \tfrac{3}{9} = \tfrac{4}{12}$

So each fraction can be converted into twelfths:

$$\tfrac{1}{4} - \tfrac{1}{3} = \tfrac{3}{12} - \tfrac{4}{12} = \tfrac{^-1}{12}$$

Now don't panic! This is not wrong but some manipulation is still needed to get the right answer. As before, we add the whole number answer and the fraction.

Step three   So the final answer is: $8 + {}^-\tfrac{1}{12} = 7\tfrac{11}{12}$

Practice questions

1. Work out the family of fractions for the fraction $\tfrac{5}{9}$ for the first six places.

2. Work out $3\tfrac{1}{2} + 4\tfrac{1}{8}$

3. Work out $6\tfrac{1}{4} - 3\tfrac{1}{3}$

4. Work out $7\tfrac{1}{2} + 2\tfrac{1}{3} - \tfrac{3}{5}$

5. Work out $8\tfrac{3}{4} + 9\tfrac{2}{3} + 4\tfrac{1}{2} - 6\tfrac{2}{5}$

How to divide fractions

Dividing fractions worries many adults, because they were originally taught to do it routinely, without understanding the mathematics involved. Let's think about a straightforward calculation:

$$1 \div \tfrac{1}{2}$$

What does this mathematical statement mean? It means, how many $\frac{1}{2}$s are there in 1 whole? The answer is 2.

Now apply the same reasoning to this calculation:

$$\frac{1}{2} \div \frac{1}{4}$$

In doing this calculation you are working out how many quarters there are in $\frac{1}{2}$.

Step one   Rewrite the question with the last fraction turned upside down and then multiply

$$\frac{1}{2} \times \frac{4}{1} = \frac{4}{2} = 2$$

In other words there are 2 quarters in $\frac{1}{2}$.

But why does this work ?

Let's look at the underlying mathematics. If you were asked to divide a £1 by 2, most people would quickly say the answer is 50p. There is another way of working out exactly the same sum: instead of dividing by 2, you could multiply by $\frac{1}{2}$. In other words $\frac{1}{2} \times 100$ pence also makes 50p.

2 and $\frac{1}{2}$ have a special relationship with each other. They are called **reciprocals** of each other. These are numbers which when multiplied together give an answer of 1.

Dividing by a number has the same effect as *multiplying by the reciprocal* of that number.

Try some yourself. Work out $300 \div 6$ and then work out $300 \times \frac{1}{6}$. What do you notice? You should see that they both give an answer of 50.

Fractions of quantities

A thing we often need to do in everyday life is to work out fractions of quantities. For example, find $\frac{3}{4}$ of 100. These calculations are actually quite straightforward, and many of them can be done quickly in your head.

Just remember that, whatever the fraction is, the denominator of the fraction tells you what you are to divide by. In the case of $\frac{3}{4}$ of 100:

Step one   Mentally work out $100 \div 4$
          This gives an answer of 25. Think about it,
          $4 \times 25 = 100$.

Step two   Multiply this answer by the numerator of the fraction.
          In other words 25×3 gives 75, so $\frac{3}{4}$ of 100 is 75.

Let's try another example

$$\frac{4}{5} \text{ of } 90$$

Whenever you have the word 'of' in mathematics, it means multiply. For example:

$$\frac{4}{5} \text{ of } 90 \text{ means } \frac{4}{5} \times 90$$

Proceed as before:

Step one   Work out $90 \div 5 = 18$

Step two   Work out $4 \times 18 = 72$, so $\frac{4}{5}$ of $90 = 72$

Practice questions

1. Find $\frac{3}{4}$ of 500.

2. What is $\frac{4}{5}$ of 200?

3. Find $\frac{2}{3}$ of £150.

4. Find $\frac{1}{3}$ of £180.

5. In a UK local council election, $\frac{1}{10}$ of the people voted Tory, $\frac{1}{5}$ voted Liberal Democrat, and $\frac{1}{2}$ voted Labour. The rest did not vote. If there are 45 000 people living in the town, how many did not vote?

Tutorial

Progress questions

1. Work out the family of fractions for $\frac{2}{3}$.

2. Add the following and then check your answers on a calculator:
   (a) $\frac{1}{3} + \frac{1}{8}$ (b) $1\frac{1}{2} + 12\frac{1}{8}$ (c) $3\frac{3}{4} + 5\frac{1}{2}$

3. Subtract the following and then check your answers on a calculator:
   (a) $\frac{3}{4} - \frac{1}{2}$ (b) $\frac{5}{6} - \frac{2}{3}$ (c) $5\frac{1}{2} - 2\frac{1}{4}$

4. Multiply the following and then check your answers on a calculator:
   (a) $\frac{3}{4} \times \frac{1}{2}$ (b) $\frac{5}{9} \times \frac{2}{3}$ (c) $3\frac{1}{8} \times 4\frac{2}{3}$

5. Divide each of the following and then check your answers on a calculator
   (a) $\frac{3}{4} \div \frac{1}{8}$ (b) $\frac{1}{4} \div \frac{1}{16}$ (c) $3\frac{1}{2} \div \frac{5}{6}$

6. Martha inherits $\frac{1}{2}$ of her mother's money. Martha's sister Mary inherits $\frac{1}{3}$ of the money. The rest goes to charity. What fraction of the money goes to charity?

Seminar discussion

1. Which is the greater, a 'half of a whole' or the 'whole of a half'?

2. 'Fractions of quantities are exact determinants of the magnitude of the quantity, whereas decimals may not be so.' Is this true?

Practical assignments

1. Find out what the Rhind Papyrus is and where it was found. Use reference books, and the internet.

2. Find out about Al Kwarizimi and how he treated fractions.

3. Examine your take home pay and determine, in terms of

fractions, what proportion of your income is spent on housing, food, books, entertainment etc.

Study and revision tips

1. Practise the techniques mentioned in the chapter and learn how to use your calculator to work with fractions.

2. Practise using calculations during shopping, particularly at the deli in the supermarket. Look at the prices of cooked meat, calculate how much per kilogram they are (they are usually given as a price per 100 grams, and there are 1000 grams in a kilogram, i.e. 100 grams is $\frac{1}{10}$ of a kilogram).

3

Percentages

One-minute overview — Percentages, like fractions and decimals, often cause adults anxiety. They feel that they should be able to work with percentages and use them in everyday life, but find them quite difficult. If that is true for you, you can safely stop worrying now that you have arrived in this chapter. You will quickly learn:

▶ the meaning of the word percentage

▶ how to work out percentages of quantities

▶ how to find one quantity as a percentage of another quantity

▶ how to increase or decrease an amount by a percentage

▶ we will also look at VAT and inflation.

What do the words *per cent* mean?

The word **cent** means a hundred. Think of a **cent**ury as a hundred years, or a hundred runs in cricket. Or again, think of dollars and **cents,** the currency of the USA where 100 cents make up one dollar, or the European currency of euros and **cents,** where 100 cents make a euro. In other words, per cent means simply 'per 100'.

So 100% of something means the whole thing. If I have one glove, I have 100% of a quantity. If I then have 2 gloves, I still only have 100%, so 100% means the whole quantity, whatever that quantity happens to be.

Gardeners often have car trailers, to carry garden rubbish. If my trailer is full, it is 100% full. My neighbour has a larger trailer than mine, but when his trailer is full, it is still only 100% full. So 100% means the whole thing.

It follows from this that when my car trailer is half full, it is 50% full. My neighbour's car trailer, when only half full, is also 50% full — even though it contains more rubbish than mine (he has a bigger garden!).

This means that 50% of his trailer load is bigger than 50% of my trailer load, but both trailers are still half full.

Calculating with percentages

As we said with fractions, there are many ways to work out percentages. In this book we have illustrated some of them. They may not be the quickest techniques but they should be easier for you to understand.

A percentage is a fraction with a denominator of 100, so percentages can easily be changed into fractions. Here are some examples:

$$50\% = \frac{50}{100} \qquad 70\% = \frac{70}{100} \qquad 90\% = \frac{90}{100}$$

You can cancel these through the $a^{b/c}$ button on your calculator. If they do not cancel, it means they are already in their lowest terms and cannot be reduced any more.

Finding a percentage of a quantity

This type of problem refers to the sort of question where you hear a statistic and want to work out how much that statistic is worth, for example 5% of £300. Here, 5% is $\frac{5}{100} \times 300$ on your calculator. The key sequence here is $5 \div 100 \times 300 = 15$, so 5% of £300 is £15. Think of percentages as pennies in the pound (or cents in the dollar/euro). In this case, it is 5p in the pound for every one of the three hundred pounds, so it is 300 lots of 5p, namely £15.

Study note

When you are calculating percentages in a money context, with a calculator, the machine will not show the last number in the display, when the number is a zero. So something like 5.6 is actually £5.60.

Another example

Take 12% of £46. Think of it as 12 pence in the pound and there are

46 individual pounds, so:

$$12 \times 46 = £5.52$$

More formally,

$$12\% \text{ is } \frac{12}{100} \times 46 = 5.52$$

You need to be able to work out the percentage in the more formal method because an examination course will often have quite strict requirements for an award; it is not worth upsetting the examiner by developing your own idiosyncratic methodology.

The reason we have included the idea of 'pennies in the pound' is to help you develop a quick method to find straightforward percentages. Remember, this more formal method is only one of a number of methods. If you have another method that you were taught at school, and you can do it well, by all means continue to use it, but make sure that you are clear and understand how your method works. The ultimate test is that your method must work every time; if it does not, we advise you to learn this method.

Some more examples
Find 60% of 450.

Step one   Write out the calculation in this manner: $\frac{60}{100} \times 450 =$

Step two   The key sequence for the calculator is:

$$60 \div 100 \times 450 = 270$$

Step three Check your answer by treating the question as if it were money. 60% is 60p in the pound, and there are 450 individual pounds. This gives an answer of 27000. Remember this is in pence, so divide by 100 and this will give you £270, confirming the earlier answer. Remember not to write down the answer as money, when the question is not actually about money.

Practice questions
1. Find 4% of 600.

2. What is 65% of 2000?

3. Martin reckons that 4% of £500 is more than 5% of £400. Is he correct?

4. Sarah receives a pay rise of 4% on her £18 000 annual salary. How much is her new salary?

5. Simon is a publishing executive. He received a budget of £80 000 last year, to develop new books. His budget is increased by 2%. How much is his new budget?

6. An insurance company offers a no claims bonus of a 35% discount on its premiums, for careful drivers. What is the reduced premium when the total premium costs £450?

7. The population of a town increased by 45% between 1950 and 1980. If the population in 1950 was 55 000 people, what was the population in 1980?

8. The price of a new car was £17 600 but is increased by 6%. What is the new price of the car?

9. A restaurant adds a charge of 10% for service. How much does a basic meal, costing £15, actually cost?

10. A new van is valued at £18 000. After one year its value has decreased by 15%. What is the van worth now?

One quantity as a percentage of another

This is a big mouthful and we need to be clear what it means, and, as we have said before, there are lots of ways of working this type of question, but we have chosen the methods that we regard as the most straightforward. The statement 'one quantity as a percentage of another' means that we need to be able to work out questions like 'what is 40 as a percentage of 120?'. These are everyday questions that you may be asked to work out in your workplace. For instance, in management of stores, what if you have to calculate the number of items left in the stores before you order new items, or if you need to calculate the days that employees have attended work, or attended training? So you can see that there are a substantial number of occasions where you must be able to answer questions of this kind. Using the example above, follow the steps that we have laid out.

Step one   Write the quantity as a ratio (fraction) $\frac{40}{120}$

Step two   Now multiply this ratio by 100

So in one go, this looks like: $\frac{40}{120} \times 100 = 33.33\%$

So 40 out of 120 is 33.3% (as a fraction, this is $\frac{1}{3}$)

Another example

Megalaw Incorporated stock widgets. At the start of the week they had 2000 widgets. On the following Saturday evening they had 300 left. What percentage of widgets was sold during the week? If they have 300 left, they must have sold 1700 during the week.
So 1700 out of 2000 is written as:

$$\frac{1700}{2000} \times 100 = 85\%$$

So they sold 85% of the stock of widgets during the week.

Let's try an example regarding attendance of employees at the place of work. Jo works at a factory; in a 30-day period she had 6 days allocated as rest days and took 4 days off for illness. In this period there were also 4 Sundays on which she did not work. What percentage of the 30 days did she actually work?

Step one   Find the number of days that she actually did work.
              30 – 6 rest days – 4 sick days – 4 Sundays = 16 days
              Therefore she actually worked 16 days.

Step two   Work out 16 as a percentage of 30
              $\frac{16}{30} \times 100 = 53.3\%$

So she actually worked for 53.33% of the 30 days. However, this is a misleading statistic, because it suggests that she was actually absent for almost half of the time she should have been at work.

Let's look at the figures a little more closely. The whole time period was 30 days, but Jo was not really expected to work each of the 30 days. Six of those days were allocated as rest days, so the period is reduced to 24 days. In those 24 days, there were 4 Sundays, therefore reducing her time to 20 days, out of which she was absent for 4 days, whilst sick. This means that she actually worked for 16 days out of a possible 20 days, so 16 as a percentage of 20 is:

$$\tfrac{16}{20} \times 100 = 80\%$$

You can see that this is a substantial improvement on the first set of figures and shows what Jo's true attendance rate is really like. She actually attended for 80% of the time that she was required to attend and is not as work shy as the first set of figures led us to believe. We will look at manipulation of figures elsewhere in this book, but it is important to consider, when working out percentages, how accurately the figures reflect the reality of what is going on. The figure of 53.33% was not inaccurate — but was it truthful?

Increasing an amount by a percentage

Sometimes you will find that a quantity has to be increased by a certain percentage. For instance, a wagon driver delivers 400 bags of compost to a garden centre. The garden centre manager asks the driver to increase the order by 40%. How many bags will the next delivery be?

Here, the first order was 100%, in other words the delivery of 400 bags was 100%. So if we are to increase the order by 40%, we need to up the order to 140% of the original order. Notice that it is possible to have an order greater than 100%. It simply means more than the original order.

So we need to work out 140% of 400 bags — and remember, this means that the new order will be more than 400 bags. So the calculation is:

$$\tfrac{140}{100} \times 400 = 560$$

In other words, to increase the order by 40% the new delivery must contain 560 bags, or an extra 160 bags of compost.

What if the order had been increased by 55%? Again, we need to add 55% onto this order, to find the number of bags of compost represented by 155% of this compost load:

$$\tfrac{155}{100} \times 400 = 620$$

In this case, the new order would be 620 bags, an increase of 220 bags on the original order. Although these are fictional calculations, they are based on the real type of calculation that does go on in business every day, up and down the land.

Decreasing an amount by a percentage

This is the opposite of what we have just been talking about. Imagine that the garden centre manager had actually asked for the driver to *reduce* the next order for bags of compost, say by 40%. In this case, if the driver is to reduce the order by 40%, his next order must be 60% of the original order. Let's consider that one again. He is asked to cut the order down by 40%. Since 100% is the whole load, he will have to bring 60% of this load for the next delivery:

$$100 - 40 = 60\%$$

So the calculation is now: what is 60% of 400 bags?

$$\tfrac{60}{100} \times 400 = 240 \text{ bags}$$

This means the next delivery will be 240 bags of compost.

What if the driver had been asked to reduce it by 55%? In this case $(100 - 55 = 45)$ he would need to work out 45% of the original load:

$$\tfrac{45}{100} \times 400 = 180 \text{ bags}$$

So the new delivery would be 180 bags.

Overview of methods

If you are increasing the percentage, simply add the increase on to 100% and then work out the percentage of the amount. For instance, if the increase is 45%, then work out 145% of the original.

If you are decreasing an amount by a percentage, simply subtract that percentage from 100%. This will then leave you with the percentage that you require. For instance, if the decrease is 35%, then subtract 35 from 100 $(100 - 35 = 65)$ so you need to work out 65% of the quantity.

VAT

VAT stands for value added tax. It is added onto the price of certain goods, as a percentage of the sale price. In other words, if you buy something priced at £10 including VAT, then not all of that amount of money actually goes to the seller, because s/he is really collecting tax on behalf of the government. VAT is currently 20% of the goods price, so 20% of that £10 is paid to the Inland Revenue in the UK. Other countries in Europe also charge VAT on their goods, but the rates vary.

Certain goods are **exempt** from VAT, such children's clothes and books. There is also a class of goods that are not exempt but on which no VAT is payable, for example, meat. Meat is **zero rated**. This means that whilst there is no VAT payable on your weekly joint, all it takes is for the current Chancellor of the Exchequer to put up the rate of VAT on meat and we will have to pay tax on the Sunday roast. (Any Chancellor who put VAT on food would be politically stupid!)

The VAT rate

VAT is currently levied at 20% in the UK, which means that the cost of the goods is increased by 20% to the customer. (If the customer is another business, and is registered for VAT, they can claim this VAT back.) There are certain levels of turnover at which all businesses must register for VAT. If you exceed this level and do not register, you can be fined. For more details on VAT registration, contact your local Revenue and Customs Office.

Some prices are quoted exclusive of VAT. That means the VAT is not included in the price and so will be added on at the check out. Prices that are quoted inclusive of VAT are those where the VAT is included in the selling price.

Let's look at some examples. A shirt costing £10 exclusive of VAT means that VAT has to be added on, so we need to find 20% of the £10, in the same manner as earlier:

$$\tfrac{20}{100} \times 10 = £2$$

So the shirt actually sells for £12, of which the VAT man gets £2.

Another way of working out the price of the shirt is — as we did earlier — to work out 120% of the price (see the examples on compost delivery, above):

$$\frac{120}{100} \times 10 = £12$$

This is an easier method, as it only takes one step.

Let's look at another example. What is the final selling price of a pack of printer paper costing £5.99? Here £5.99 is 100% of the price and we have to add on 20% VAT:

$$\frac{120}{100} \times 5.99 = 1.198. \text{ To the nearest penny this is } £1.20.$$

So the cost of the paper is actually 5.99 + 1.20 = £7.19, quite a substantial increase.

An easier way would simply be to work out 120% of 5.99, and do the calculation in one step:

$$\frac{120}{100} \times 5.99 = £7.188. \text{ To the nearest penny this is } £7.19.$$

How to calculate the actual price when VAT is included

Sometimes when the price is listed as containing the VAT, it is desirable to be able to unravel the cost of the commodity and to work out how much was the actual cost of the goods and how much was VAT.

Example

If I buy a beautiful statue for my garden, costing £80, inclusive of VAT, how much was the VAT?

Here the £80 represents the cost of the garden statue and the VAT, so it must represent 120% of the actual cost. Let's just think about this for the moment. The cost (£80) is made up of 100% of the statue cost, and 20% represents the cost of VAT.

Therefore, £80 represents 120%. If we turn this around, we can simply say that 120% is represented by £80, or is **equivalent** to £80. There is a mathematical symbol for **equivalence**, namely ≈, so we can say:

$$120\% \approx £80$$

Notice, we have not used the equal sign. These two quantities are *equivalent* but strictly speaking in mathematical terms, they are not *equal*.

Since 120% ≈ £80, we can work out what 1% is worth, and therefore we can work out what 100% is worth:

$$120\% \approx £80$$

$$\frac{120}{120} \approx \frac{£80}{120}$$

Here we have divided both sides by 120, because this reduces the left hand side to 1%. Remember, whatever we do to one side, we must do exactly the same to the other side.

$$1\% \approx 0.666666666$$

so 1% is worth just over 66 pence. This means that 100% must be

$$100 \times 0.666666666 = £66.67$$

To the nearest penny, the statue cost £66.67. The remaining £13.33 was paid to the government in taxation.

Let's try another example. Imagine that you are having a swimming pool fitted in your garden at a cost of £15 000 inclusive of VAT. How much is the cost of the pool, *without* the VAT? As in the previous example, we set up an equation where the total cost of the pool inclusive of VAT is equivalent to 120%:

$$120\% \approx 15\,000$$

Now divide both sides by 120, to find what 1% is worth

$$\frac{120}{120} \approx \frac{15\,000}{120}$$

$$1\% \approx 125$$

so 100% ≈ £12500. The remaining £2500 is VAT that is paid to the government.

What is inflation?

Inflation is the rise in price of goods and services, throughout the economy. In simple terms, this means that goods and services are more expensive now than a year ago. Economists differ in explaining the cause of inflation, according to their political leanings. The main point is that inflation means prices are going up.

Even when you hear that 'inflation is down', it does not mean that prices are coming down. It means that prices are still going up, but *not as fast*. In other words it is the **rate of increase** that has fallen, not the prices themselves.

Suppose inflation was 4% but has now fallen to 2%. It means that prices were increasing by 4% but that rate of increase has now fallen to 2%. But the prices are still actually going up.

Pay rises

Because of inflation, the unions have traditionally demanded an increase in pay that at least keeps pace with inflation. They argue that if a workers earn £100 a week and the rate of inflation is 2%, then they need an increase in pay of 2%, simply so that their buying power stays the same as it was last year. To receive a pay rise that is less than the rate of inflation is, in some people's eyes, actually a pay cut.

Calculating your pay rise

This is the same type of problem as we mentioned earlier. Take the top line of your pay slip (the gross pay). Using the year-on-year inflation rate, find the new level of salary/wage that you need to achieve, in order to 'stand still' economically.

Let's say Mike works in a factory for £200 a week. If the year-on-year inflation rate is 4%, what does Mike need to earn next year, in order to have the same buying power? Answer: his new pay needs to be 104% of his old pay:

$$\frac{104}{100} \times 200 = £208$$

It means that, to keep the same buying power, Mike needs to have his pay increased to £208 a week.

What is a ratio?

A ratio is a relationship between one quantity and another. We use ratios in many aspects of life, for example mixing concrete, in compost mixing in the garden and in cooking. When you hear a builder talk of two parts sand to one part cement, he is talking in ratios. It may be that those parts are shovelfuls, or kilograms — it doesn't matter, they are still ratios.

Suppose the ratio was two parts sand to one part cement. It simply means that for whatever amount of sand that you put into the mix, you always put half that amount of cement.

For example, 4 shovels of sand would mix with 2 shovels of cement, whereas 8 shovels of sand would mix with 4 shovels of cement. In other words, the amount of sand is always twice the amount of cement.

Notice, in mathematics, a ratio is written using a colon. For instance, the ratio 3 to 1 is written as 3:1.

Solving problems using ratios

When solving a problem, it is vital to know the ratio and the total amount or quantity in the problem.

Imagine we had a compost mix for a lawn top dressing. It is made up of earthworm casts and sharp sand, in a three to one mix. There is a total of 10 000 kg of compost. How much earthworm casts was used?

Remember, the mix was 3:1 (three to one). In other words there is three times as much earthworm casts in the top dressing as there is sand. In this case, the top dressing is made up of 4 parts all together, so we can work out the value of one part:

$$\frac{10000}{4} = 2500$$

So each part of the 10 000 kg mix is 2500 kg. Since the ratio was 3:1, the amount of earthworm casts used was 3 × 2500 = 7500 kg. The remaining 2500 kg was made up of sand.

Another example of ratios

Concrete is mixed in the ratio of three parts fine sand to one part sharp sand to two parts cement. How much cement is used to mix 500 kilograms of concrete?

First of all, rewrite the ratio in a more friendly form, like this: 3:2:1. This stands for three parts fine sand to two parts cement to one part sharp sand. Altogether this makes 6 parts (3 + 2 + 1). So 6 parts is equivalent to 500 kg of concrete. You write this mathematically:

$$6 \approx 500$$

so we can work out what *one* part is worth:

Now divide both sides by 6:

$$\frac{6}{6} = \frac{500}{6}$$

$$1 \text{ part} = 83\tfrac{1}{3}$$

So two parts of cement in 500 kg of concrete must be two lots of $83\tfrac{1}{3}$, namely $166\tfrac{2}{3}$ kg.

Since there are 1000 grams in a kilogram, $\tfrac{2}{3}$ of a kilogram is 666.6 grams. The total amount of cement needed in 500 kilograms of concrete is therefore 166.66 kg.

What is proportion?

Two quantities are said to be in proportion if they increase or decrease by the same ratio.

For example, suppose potatoes are 20 pence per pound and I buy two pounds. I expect to pay 40 pence. I would also expect to pay 10 pence for $\tfrac{1}{2}$ lb (half a pound weight).

Income tax is a good example of proportion. The more you earn, the more income tax you pay. The *proportion* of your income that goes in tax is the same as other people's, but the *amount* depends on your income.

It is not quite as simple as this example suggests, because there are bands of income which are taxed at different rates. However, for the sake of simplicity let's say it is a straightforward 20%.

For example, Fred earns £15 000 and Roger earns £95 000. How much tax do they pay at 20% ?

If all of their income was taxed at 20%, Fred would pay

$$\frac{20}{100} \times 15\,000 = £3000$$

and Roger would pay:

$$\frac{20}{100} \times 95\,000 = £19\,000$$

However, there are tax bands. This means that your income is treated as if it is made up of components. To find out the current tax rates, contact your local Inland Revenue office. Here we have made up some rates, to make it clear. Let's imagine Roger, a married man, is a high earner on £200 000 a year.

The first part of his income is made up of a tax free part (called the allowance). For the sake of this example, let's say Roger's allowance is £7475. Therefore his taxable income is

$$£200\,000 - £7475 = £192\,525$$

His rate of tax is 20% on the first band of £3500, so on this part of his income he pays:

$$\frac{20}{100} \times 3500 = 7000$$

The next rate of tax is 40% on the next band of 35k to 150k:

$$\frac{40}{100} \times 11\,500 = £4600$$

So his total tax bill is £7000 + £46 000 + £21262.50 = £74 262.50 but we are not finished yet. Remember Roger has £192 525 of taxable income so we now tax the part between £192 525 and £150 000. This is £42 525 and is taxed at 50% so $\frac{50}{100} \times 42\,525 = £21\,262.50$.

In reality, Roger would have the right to claim tax allowances against certain items that he may have bought and would therefore be able to decrease his tax liability. However, this fictional example does show how your income tax liability is in proportion to your earnings.

Practice questions

1. Find £40 as a percentage of £500.

2. Increase the sum of £20 000 by 35%.

3. Decrease the sum of £15 000 by 9%.

4. A coat is sold in a sale, with a 20% discount. If the pre-sale price of the coat was £160, how much is the coat in the sale?

5. Work out the VAT payable on 5 litres of paint that costs £1.20 per litre.

6. The price of a piece of timber in a DIY store, inclusive of VAT, is £7.50. How much of this price is VAT?

7. Explain, in mathematical terms, why three first-class stamps are three times as expensive as one first class stamp.

8. If the year-on-year inflation rate is 4%, and Steve earned £26 000 last year, what amount of pay rise should he request, in order to keep the same standard of living?

9. Two people share an inheritance in the ratio of their ages. The inheritance is £400 000. One of the people is 40 years old, the other is 60 years old. How much does each person get?

10. A and B invest money in a business. A invests £8000, B invests £12 000. They agree to split the profit in the same ratio as their investment. At the end of the year, the profit is £50 000. How much does each person get?

Tutorial

Progress questions

1. Find 14% of 900.

2. Increase 1400 by 65%.

3. Decrease 800 by 12%.

4. A tool costs £12 + VAT. At 20%, how much VAT is payable?

5. A tin of paint, inclusive of VAT, costs £9. If the VAT is rated at 20%, how much of the cost is VAT?

Seminar discussion

1. A trade union leader, when discussing percentage pay rises, once famously said that percentages give 'Most to those who need it least, and least to those who need it most'. Was he right?

2. 'If you invest 10% of everything you earn, and do not touch it, you will retire a millionaire.' Do you think this is true?

Practical assignment

Imagine you are buying a vacuum cleaner. Vacuum cleaners can be bought for £150 + VAT at 20%, but there is a discount of 8%. From your point of view, as the customer, is it better to knock the discount off the £150 and then add the VAT, or better to add the VAT and then knock off the discount?

Study and revision tips

1. Always think of percentages as pennies in the pound, so 10% is the same as 10p in the pound.

2. In an examination, even if you are using a calculator, always show clear working. If you make an error, the examiner will be able to see your reasoning and may award marks, even though the actual answer is wrong.

3. Read your calculator booklet and make sure you understand how to use its functions.

Working with Decimals

One-minute overview — Decimals cause anxiety for many adults, but with a little concentration, it soon becomes clear. Decimals are also known as decimal fractions. They have only been used in Europe for a few hundred years. It is important that you feel comfortable and confident in dealing with decimals. In this chapter you will study:

▶ the meaning of the different place values in decimals

▶ how to change decimals to vulgar fractions

▶ how to round off in decimal places

▶ working with a decimal of money

▶ using a calculator when dealing with decimal money systems.

The meaning of place value

Place value is a term that refers to the value of a number when it occupies a certain position in a figure. We discussed this earlier in the book, but it is worth mentioning again here.

1. If you have a number like 375, how much is the 7 worth? Since it is in the 'tens' column, it is worth 7 lots of ten, in other words 70.

2. Similarly, the 3 is in the hundreds column. Therefore it is worth 3 lots of a hundred, or 300.

3. Since the 5 is in the units column, it is worth 5 units, or 5 lots of one, which is just 5.

Remember, numbers after the decimal point are expressed individually. For example 0.35 is said as 'nought point three five'. It is wrong to say this number as 'nought point thirty five'.

In the same way, decimals are placed in columns that signify the value of that column. The decimal part of a number is made up of the decimal point and the numbers to the right of the decimal point.

decimal point | tenths | hundredths | thousandths |

Notice how the decimal point is to the left of the tenths column. You should always think of the decimal point as being static, and that the numbers move around the decimal point.

The tenths column

This is the decimal equivalent to dividing by 10. In other words, 0.1 and $\frac{1}{10}$ are equal. Therefore any number in the tenths column is the same as a fraction with a denominator of 10. For example:

$$0.3 = \tfrac{3}{10} \quad 0.7 = \tfrac{7}{10} \quad 0.5 = \tfrac{5}{10} \quad \text{and so on.}$$

Notice that when the number in the units column is zero, we put a zero in the column. You can write decimal numbers without the zero, so 0.25 becomes .25 and is expressed as 'point two five'. However, you are strongly advised to place a zero in the units column, and avoid expressing decimals as the decimal point followed by the decimal.

People are known to read numbers wrongly, if there is no zero in the units column. In one reputed case in the late 1960s, a nurse is said to have misread a decimal number and injected a baby with ten times the amount that she should have done; the baby is said to have died as a result of the mistake. We cannot vouch for the accuracy of this anecdote but it shows the importance of getting it right.

The hundredths column

The hundredths column is the second column after the decimal point. It equates with fractions that have a denominator of 100, for example,

$$0.03 = \tfrac{3}{100} \quad\quad 0.07 = \tfrac{7}{100}$$

As you can see, the numbers in this column are rather small.

The thousandths column

Numbers in the thousandths column are even smaller, very small indeed. For example,

$$0.007 = \tfrac{7}{1000} \qquad 0.009 = \tfrac{9}{1000}$$

How to change decimal fractions to vulgar fractions

Just as a reminder:

▶ a **vulgar fraction** is a fraction like $\tfrac{1}{2}$ or $\tfrac{3}{4}$

▶ a **mixed number** is one like $1\tfrac{1}{2}$ which is made up of both a whole number and a fraction

▶ an **improper fraction** is something like $\tfrac{15}{3}$, where the numerator is larger than the denominator.

Let's take an example of 0.3. Since the 3 is in the tenths column, then this must be $\tfrac{3}{10}$. Notice, 0.3 is the same as 0.30 or 0.300 or 0.3000000 and so on. In fact you can have as many zeros as you like after the three, they do not affect the value of the 3. For example:

$$0.3 = 0.3000000000000000000000000000000000000$$

Both numbers have 3 in their tenths column.

We always use the column at the end of the decimal number (provided it is a non-zero figure) to show the value of the fraction. In a number like 0.35, the 5 is the last number and is in the hundredths column. Therefore the number as a fraction is

$$\tfrac{35}{100}$$

If you think about it, this makes good sense. Since the 3 is in the tenths column, it is clearly $\tfrac{3}{10}$. But from our 'family of fractions' work, you should see that $\tfrac{3}{10}$ is also the same as $\tfrac{30}{100}$ and since the 5 is in the hundredths column, or $\tfrac{5}{100}$, then

$$\frac{30}{100} + \frac{5}{100} = \frac{35}{100}$$

Cancelling fractions

If this is something that causes you a problem, fear no more, because the ab/c button on your calculator can help. Let's say you want to change the decimal 0.5 to a fraction. You need to know that it is a fraction in its lowest terms. The decimal number 0.5 has a 5 in the tenths column and so becomes $\frac{5}{10}$. But if you put $\frac{5}{10}$ through your calculator, you will find that it changes. Try it — what do you get? You should have found that $\frac{5}{10}$ is the same as $\frac{1}{2}$.

Without a calculator, just look for the largest number you can divide exactly into the numerator and the denominator of the fraction. In our example, that number is 5. It goes into the numerator once and into the denominator twice, hence $\frac{5}{10} = \frac{1}{2}$.

Changing vulgar fractions to decimals

This is also quite straightforward, but you will need to think carefully about what you are doing. Let's take a fraction like $\frac{1}{2}$. This means 1 part, split or divided into 2 parts. To change this into a decimal, we need to divide the numerator by the denominator. Since a decimal $\frac{1}{2}$ must be '0.something', because $\frac{1}{2}$ is less than one, so its decimal equivalent must be less than 1.

$$\frac{1.0}{2}$$

We can write 1 as 1.0 because it does not affect the value of the 1. Now divide the 2 into the 1.0 (treat it as 1.0 ÷ 2). This gives you an answer of 0.5.

Confirm this by using your calculator. Use it to divide the numerator by the denominator, so key in 1 ÷ 2 and you should find this confirms what we have already found.

Now try this one. Change $\frac{1}{4}$ into a decimal as before. Divide 1 by 4, so write it as:

$$\frac{1.0}{4}$$

Again, treat the 1 as 1.0 and divide this by 4:

$$\frac{1.0}{4} = 0.2$$

This gives a remainder of 2. In fact the remainder is 0.2, so we can justifiably put a zero behind it:

$$\frac{1.0^{r2}}{4} = 0.2$$

Now divide 4 into this remaindered 0.2, giving you the answer 0.05. So $\frac{1}{4}$ is **0.25** as a decimal. Again, confirm this on your calculator.

Rounding to decimal places

Using the correct symbols
Rounding means approximating, or giving an answer that is pretty close to the actual answer. The point here is that a rounded or approximated answer is not exactly the same as the actual answer.

For example, an answer of 100.2 is 'approximately equal' to an answer of 100 but it is mathematically wrong to write it as 100.2 = 100 because they are not actually equal.

In mathematics there is a useful symbol that means 'approximately equal'. It is:

$$\approx$$

So it is quite legitimate to write

$$100.2 \approx 100$$

since this means they are approximately equal.

How to round decimals
Imagine a number line, going from one number to another. Take for example the numbers 1 and 2. We could draw it like this:

1_____2

In fact, there are a whole range of numbers between 1 and 2. They include numbers like 1.25, 1.5, and 1.75, and all points in between these numbers.

1_____1.5_____2

This gap between the two main numbers is called an **interval**. The rule is: if the value you wish to round is halfway or more, then you round up to the *higher* value of the interval. If the value you wish to round is less than half of the interval, you should round *down* to the lower value of the interval.

For example, round 1.6 to the nearest whole number. If we show this on an interval, it would look like this:

1_____1.6_____2

Since 1.6 is more than half of the interval, it should be rounded up to the higher value of the interval. So in this case, 1.6 rounds to 2. Mathematically we write this as:

$$1.6 \approx 2$$

This makes good sense, because if you look at the interval, you can see that 1.6 is actually physically closer to 2 than it is to 1.

What about smaller intervals?

These are treated in exactly the same way. For instance, round 1.87 to the nearest point 1. If we represent 1.87 on a number line, it appears in the interval between 1.8 and 1.9:

1.8_____1.87____1.9

In this case, we want to round to the nearest point 1. This means that the answer will be either 1.8 or 1.9. To determine which of these is the correct answer, we look at the next number after the first decimal place. In this case it is 7. Since 7 is more than half way in this interval, we must round up 1.87 to 1.9.

In this example, we have just rounded 1.87 to one decimal place. This means that 1.9 is an approximation (and a very good one) to 1.87. If you think about it, 1.9 is only 0.03 away from 1.87.

From the work we have done earlier, we know that 0.03 is the same as $\frac{3}{100}$. Our approximation in the example above is actually only $\frac{3}{100}$ away from the correct value: this is clearly a very good, close approximation.

What about approximating to two decimal places?

Approximating to two decimal places means that the final answer will be an approximation with *two* numbers after the decimal point. Therefore, we need to examine the third number after the decimal point. That will help us to determine the approximation.

For example, suppose we had to round 2.784 to two decimal places. What is the third figure after the decimal point? The standard rule is that if it is 5, or more than 5, we must round the previous figure up one. If it is less than 5, we must leave the previous figure alone. Since the previous figure is less than 5, we simply approximate the number as 2.78.

When you are approximating and giving an answer as an approximation, you *must* always state the approximation. You should usually state it in brackets after your final answer.

In this case, the approximation would be 2.78 (2 d.p). The 2 d.p. in brackets indicates the accuracy that you have rounded to, in the question.

Do I round in the middle of questions?

Generally speaking, the answer is no. By 'rounding in questions', we mean approximating earlier on in a calculation, and then going on and using that approximation further into the calculation. This builds in what are called **rounding errors**.

Suppose I had to work out the volume of a cylinder, say a tin of beans, and had been told the height and the radius of the end of the tin. Let's say the radius of the circle on the end was 5 cm and the height was 10 cm. Now let's work out the volume. From your school days, you may remember that the cylinder is a **prism** (a solid of uniform cross-section). Therefore the volume of the prism is the area of the cross section × the length of the prism.

In the case of cylinders, there is a formula, $V = \pi r^2 h$. So I need to find the area of the circle on the end of the cylinder. I could actually do this in one step, but for the purposes of this example, we will do it in two steps:

$$A = \pi r^2$$
$$= \pi \times r^2$$
$$= \pi \times 5^2$$

In school you may have been told to take π as 3.14, but from now on, you should always take the value given by your calculator:

$$A = 78.53981634 \text{ cm}^2$$

Don't forget the squared units: it is area. Using $\pi = 3$, an approximate value, gives us

$$A = 75 \text{ cm}^2$$

If I now use the 75 cm^2 answer in finding the volume of the cylinder, I am using an approximation, so the answer will not be as accurate as it could be.

Look at this: using the area of 78.53981634, the volume is

$$78.53981634 \times 10 = \mathbf{785.3981634} \text{ cm}^3$$

which rounds to 785.54 cm^3. But if we use the rounded area 75 the volume is 75 × 10 = **750** cm^3. This is quite a big difference. This is what happens when rounding errors creep into a calculation, so do try to avoid it as much as possible.

Dealing in money

Money uses the decimal system. The word **decimal** actually means numbered in tens or proceeding by tens. The current British system of 100 pence = £1 is a decimal system. In the USA it is similar, based on 100 cents = $1. In many countries in Europe the currency is the Euro, and €1 = 100 cents. It is worth noting that different countries place the € sign in different places, e.g. €10 or 10€. It is still the same value.

Avoidable mistake 1

In the current British currency, many people write sums of money in pounds and pence by using both the £ sign and the p sign, for example £3.01p. This is wrong. Whenever you write in pounds, you should not use the p for pence sign. To indicate three pounds and one penny correctly, write it as £3.01

If you want to write a sum like this in pence, just remember that there are 100 pence in £1, so write it as 301p. In this case it is correct to have the pence sign, p. In short, remember not to use the £ sign and the pence sign together.

Avoidable mistake 2

When dealing in any calculations involving money, you must always work to 2 decimal places. Why? Because the currency always needs 100 units to make up the whole, for example 100p = £1, and 100 cents = $1 and 100 cents = €1.

This is particularly important if you are using a calculator. Try this on a calculator: £3.60 + £4.80, for example. What answer did you get? Your calculator display will probably say 8.4. But what does this 8.4 actually mean? Think about what you now know about decimal places: the 4 is in the tenths column, so it is worth 40 pence. Consequently, a calculator display of 8.4 means £8.40.

What about pounds and single amounts of pence?

What would the calculator display if the answer was eight pounds and four pence? Again, let's think of our work on decimal places. The four pence needs to be in the hundredths column, since 4 pence is $\frac{4}{100}$ of a pound. This means it has to go in the second column after the decimal point, so eight pounds four pence on a calculator display will look like 8.04.

The Euro and Britain

At the time of writing the UK has kept its own currency and it not intending to join with many European countries and introduce the Euro. There are many political and economic arguments to be had over the Euro but the fact is, it does exist. This means that regardless of whether Britain stays outside of the Euro or not, we do need to understand it and how it impacts on our lives. It is the case than many businesses, particularly those near ports, now run two cash tills, one for Pounds and one for Euros. It is also the case that a substantial number of UK companies, Aber Publishing (GLMP Ltd) included, are now obliged to price their products in both Pounds and Euros.

This does mean extra costs for business in that they may have to keep two sets of accounting. However there are a number of UK residents who actively want the UK to pull out of Europe. This is a political decision and outside the scope of this book but we wish to make it clear that the Euro does and will continue to impact on the lives of all UK residents, even if only at the level of currency exchange for holidays.

How does the euro work?

Quite simply, the euro is the face value of the note, and it is made up of cents. There are 100 cents to one euro. At the time of writing, the rate of exchange between the pound sterling and the euro is roughly that one euro is worth about 70p. Until the UK decides to join the European Monetary Union, we will all have to learn how to convert euros to pounds.

For example, how much is £3.50 in euros? First of all, check the exchange rate. On the day we did this, the euro was worth 70p, so:

$$\frac{£3.50}{70p} = 5 \text{ euros}$$

The symbol for the euro is €. It is already on some of the newer keyboards, but you can download it from the Microsoft web site. The symbol is placed in front of the currency in most countries, so 5 euros is written as €5. However, in countries like France where the French franc symbol came after the currency, they now write five euros as 5€.

Tutorial

Progress questions

1. Write down the value of the underlined figure in each of the following:

   (a) 32<u>5</u>   (b) 1<u>7</u>89   (c) 3<u>2</u>5 459

2. Change 0.45 to a vulgar fraction.

3. Round each of the following to 2 decimal places:

   (a) 20.456   (b) 49.5879   (c) 127 456.2345

4. Change $\frac{3}{8}$ into a decimal.

Seminar discussion

1. 'The problem with decimals is that they can sometimes only be approximate values, whereas vulgar fractions can always be determined exactly'. Do you agree?

Practical assignments

1. Visit the UK Treasury web site at http://www.euro.gov.uk for an update on the situation regarding the (€).

2. Imagine that you were leaving the UK and travelling through Europe. If you had to pay 10% commission to change your currency, as you passed through other countries, how much would you have left out of an initial £100?

Study and revision tips

1. Make sure that you are familiar with the fraction button on your calculator.

2. Remember, when rounding to decimal places, to look at the next value after the required number of places, then make your decisions.

| | 5 | |

Using a Calculator

One-minute overview — In this chapter we are going to discuss and study the use of the calculator. The calculator is an essential tool in any student's case and it is a tool that can save you time. We will look at:

▶ the use of fractions on the calculator

▶ the use of the memory

▶ how to use the second function keys

▶ the meaning of the term reciprocal, and whether your calculator has a reciprocal button

▶ the meaning of the ANS button.

It is important that you spend some time and become familiar with your calculator. Using a calculator is a bit like driving a car. After you have been driving one for a while, you automatically reach out for where you know the correct buttons and switches are located. You need to be that familiar with your calculator.

Some students say that they will get the correct calculator just before their examination or assessment, and then learn how to use it. This is not advisable, and is frankly poor study technique. If you do not have a calculator, you must buy one, and preferably a scientific calculator. The reasons why you should choose a scientific calculator will become clear later in this chapter.

Why use a calculator?

The calculator is a tool that frees you from the tedium of crunching numbers. It will allows you to explore mathematics in greater depth. Readers over the age of forty will probably remember the tedium of working out things such as square roots. They will recall the agony of having to work with logarithms as an aid to calculation, without really understanding what a log or an antilog was — just having to use it, and crunch the numbers with it.

What most people regard as mathematics — adding and subtracting, mentally calculating, percentages, fractions and decimals — is in fact only a part of mathematics. Real mathematics involves applying knowledge to problem-solving situations. The media often claim that the use of calculators damages a child's mathematical development. If you are a parent, this may cause you some anxiety. It is not true. It is at best a simplistic interpretation of a very complex situation.

Some students certainly do become calculator dependent. Whatever the display on a calculator says, they simply repeat, without understanding what it means.

For instance, one student calculated 30% of 400 as 4130. This is an example of where a student fails to understand the nature of percentages; she is dependent on the calculator because she doesn't understand percentages. Removing the calculator will not help her understand how to work out percentages. It will simply hamper her mathematical development. As a student yourself, or possibly as someone with children at school, the main thing is to use the calculator properly.

No one should depend on a calculator to work out simple problems like 3×2. This is the mathematical equivalent to filling a bath full of water to wash your face. It would be ridiculous. Sensible use of the calculator frees students to study mathematics at a deeper level, and this should be encouraged.

The keyboard layout

In this section we are going to look at the layout of a scientific calculator. We will not, at this stage, look at the basic four-function calculator.

Why is there writing between the keys?

If you look at the keyboard of a calculator, you will see various bits of writing between the keys. The reason for this is that each button has more than one job or function to do. The alternative would be to have a machine with twice as many buttons; it would be at least twice the size, and rather unmanageable.

How do I operate the second use of the button?

On your keyboard you will find a button, usually at the top left or the top right of the keyboard, marked

2nd F   meaning second function, or
Inv   meaning inverse, or
Shift

You need to press this button first, before you can activate the second function of the button.

The main keys for basic maths

The fraction key

Look for a button marked $a^{b/c}$. This is the fraction key. To enter a fraction like $\frac{1}{2}$ you need to press 1, then $a^{b/c}$ and then 2. The display will look something like a one, then an L shape on its side, and then a two.

Similarly, to key in a mixed number like $1\frac{1}{2}$, simply key the first 1, then press $a^{b/c}$ and then the second 1. Press $a^{b/c}$ again, and then the 2. The display will then show the fraction $1\frac{1}{2}$.

The $^+/_-$ key

This key changes the sign of the value in the display. To enter -3, on most machines you will need to enter 3 and then press the $^+/_-$ key. You should find that the display now shows -3.

The $\sqrt{}$ key

The $\sqrt{}$ symbol means **square root**. The square root of a number is the number that you must multiply by itself to generate that first number. For example, the square root of 100 is 10, because 10 × 10 = 100. The square root of 25 is 5, because 5 × 5 = 25.

Mathematically, when you are writing 'the square root of 25', you write it as $\sqrt{25}$.

However, on the calculator, you need to experiment. Modern calculators work in the sequence that you would expect. You press the $\sqrt{}$ button first and then you key in the number that you want to find the square root of, for instance $\sqrt{36}$. This is keyed as $\sqrt{}$ and then 36 and then =.

Older machines work rather differently. You have to put in the number first, then press the   symbol. Experiment with your own machine, to see which type it is. Don't worry if it is an older machine; it will work just as well as a newer one and you certainly do not need to replace it.

The x^2 button

This is the **square** button. In mathematics, when you 'square' a number, it means that you multiply it by itself. For example 10 × 10 is mathematically said to be 'ten squared' and is written as 10^2. To key this on your calculator press the number you want to square, such as 10, and then the x^2 button. It will show you that $10^2 = 100$ (because 10 × 10 = 100). In the same way, $5^2 = 25$.

The $\frac{1}{x}$ button

This is a very useful button, because it helps to work out the **reciprocal** of a number. The reciprocal is a second number that, when multiplied with the first one, gives the answer 1. For example, the reciprocal of 2 is $\frac{1}{2}$, because 2 × ½ makes 1.

Similarly the reciprocal of $\frac{1}{5}$ is 5 because they multiply together to make 1. So, if you want the reciprocal of a number, simply key in the number and press the button. Notice that the machine will give you the answer as a decimal.

The memory buttons

The actual lettering on the memory buttons will depend on the type of calculator that you have. Some machines have a large number of memory banks. We have one, now rather old, programmable calculator, which has 64 memory banks. Usually the memory buttons are marked as M+ and MR.

▶ The M+ button can be used to input information into the memory bank. Press 6 and then M+ and you will enter the data. Now press the AC button to clear the display.

▶ The MR button is the memory recall button. If you now press it, the 6 should reappear in the display.

▶ You should also find a button marked M–. (Or it may be the second function on the M+ button.) This button clears the memory bank. If it works as a second function, you will need to press the 2nd F or INV or SHIFT button first.

The x^y button

This button is very useful. It lets you calculate any number to a **power**. For instance, 5^3 means five to the power of three: $5 \times 5 \times 5 = 125$. If you key 5 x^y 3 and then $=$, the display will give you the answer of 125.

The brackets buttons

We will examine the effects of using brackets later in this chapter. For the moment, you just need to recognise the buttons. The left hand one is usually marked as [(, and the right hand one as)].

The ANS button

Some calculators have an ANS button. It stores the previous answer that you have worked out. Suppose you are working through a calculation, say 1+2+3, and you have forgotten a previous answer. By pressing the ANS button, you can retrieve the previous answer.

The π button

This button can be found in one of several different places on the keyboard, depending on the make of machine that you own. The value that your calculator gives for π is more accurate than the traditional value of 3.14 that you were probably taught to use in school.

An investigation into different calculators

The purpose of this investigation will become clear as we move through it. First of all, you need a scientific calculator and a basic four-function calculator. On both calculators work out the following:

$$1 + 2 \times 3 =$$

What do you notice here? Can you explain what is happening and why? Does this always happen? Try other calculations like this, does the same thing happen?

It is important to understand the mathematics that is going on here. There is an order of operations in mathematics, a certain order of doing things. The two calculators have reached different answers because they work in different ways. They have different logic systems. The scientific calculator will have given you the correct answer. The statement $1 + 2 \times 3$ means $1 + 6$, so the answer is 7.

In a mathematical statement like this, you always do the multiplication first. The multiplication and division always take place before addition and subtraction. The basic four- function calculator works on a different logic system. It does the calculation in the order that it is written. So it simply works out $1 + 2$ to get an answer of 3, and then works out 3×3 to give you 9, which mathematically is wrong.

If you want to work out the calculation this way, you need the help of brackets. To work out $1 + 2 \times 3$ to get an answer of 9, bracket the $1 + 2$ part like this:

$$(1 + 2)$$

This 'tells' the calculator to work out this part first, and *then* to work out the $\times 3$ part.

More on brackets
By using brackets, you 'inform' the calculator the order in which you want the calculation to be carried out. Otherwise the calculator will work the answer out by doing multiplication and division before addition and subtraction.

Try working out $(1 + 2) \times (4 + 5)$ in one go, using your calculator. You should get an answer of 27; if not, do it again and make sure that you key carefully.

What about double brackets?

Let's say you have to work out three multiplied by four first, then add five to this answer and multiply it all by 10. How would you write all this down?

The answer is: you need to use double brackets. For this purpose we usually use square brackets, like this [ ], followed by the usual round brackets ( ). So this question is written as:

$$[((3 \times 4) + 5) \times 10]$$

To check you have written the brackets correctly, make sure you have an even number of brackets. Here the two square brackets link, then the first round bracket in front of the 3 matches with the second round bracket after the one, whilst the second bracket, before the 3, matches with the bracket after the 4.

What is the DRG button ?

Most students doing basic maths need to have their calculator working in **degree mode**. Your keyboard will have some type of **degrees, radians**, or **gradians** button. It may be as a second function or in a mode function. Read your calculator booklet carefully to see how your machine works.

You will almost certainly have to work with degrees. Radians are an alternative way of measuring angles and are used mainly at post-16 level and beyond. Gradians are another specialist measurement technique and you will not usually need to use this function. Make sure you are in degree mode. If you calculate in radian or gradian mode, you will get the wrong answer.

What is a graphical calculator?

A graphical calculator is a hand held calculator that can draw graphs for mathematicians and scientists. They are not particularly cheap. Unless your course tutor says that you need one, or you have a source which is not too costly, we suggest that you save your money. Students who are studying maths and/or science at post-16 level may find one extremely useful, but most other students can manage with an ordinary scientific machine.

Graphical calculators can be linked with PCs and perform many surprising functions. For instance, you can place a sensor on a graphical calculator and bounce a drink can on an elastic band near the sensor. The machine will draw a graph of the movement of the can, on its screen, while the can is actually bouncing on the elastic band. There are many existing programs that will fit most students' requirements, so if this does apply to you, contact the maker of your graphical calculator, via their web site. Here you should find plenty of programs and you can usually download them to your PC, for free.

Tutorial

Progress questions

1. What key sequence would you use to enter $\frac{1}{2}$ on your calculator?

2. What is the purpose of the $\sqrt{}$ button ?

3. How do you enter a number into the memory of a calculator?

4. What is the function of the ANS button on a calculator?

5. How should you enter 5 raised to the power 4, or 5^4, on your calculator?

Seminar discussion

1. 'Calculators liberate students and enable students to study at far higher levels of mathematics.' Do you agree? If so, why? If not, why not?

2. 'Calculator dependency needs to be avoided. By calculator dependency I mean the need to use a calculator to work out simple calculations like 2 × 3.' *James Craig.*

Do you agree? Is the view that James is putting forward in any way at odds with the view put forward in the first quote?

Practical assignments

1. Investigate how to use a computer spreadsheet, and how to enter formulae for calculating in a spreadsheet.

2. Set up a spreadsheet to convert from pounds sterling to euros (€).

Study and exam tips

1. Spend some time working with your calculator, and make sure that you are familiar both with the function and location of the keys. Becoming used to a calculator is very similar to becoming familiar with the location of the buttons and switches in a car. When you are familiar with the location of the buttons on your machine, it will speed your calculating.

2. Ideally, you should spend some time practising using your calculator in your non-writing hand. For instance, if you are right handed, you should practise keying with your left hand and writing with your right hand. This will speed your work, particularly in an examination.

Angles, Turning and Pythagoras

One-minute overview — Geometry is the study of lines and angles. In this chapter we are going to help you gain a basic understanding of geometry. Are you studying or working in applied studies, creative subjects such as fashion and design, or a technical subject such as engineering and building? If so, the basic geometry here will underpin your current work. An understanding of basic geometry is vital in engineering and building and in creative subjects. The Greeks developed a great deal of the geometry that we use today. In fact the great mathematician Archimedes was killed by invading Roman troops whilst he was drawing geometrical diagrams in the sand. One of the most surprising facts about the Greeks was that, despite their advanced knowledge of geometry, they did not develop other areas of mathematics to the same extent. In this chapter you will study:

▶ the nature of turning, including positive and negative turning

▶ the names of different types of angles and triangles

▶ angles on a straight line and around a point, and how to calculate missing angles

▶ parallel lines and angles

▶ finding the interior angle of a regular polygon

▶ the theorem of Pythagoras.

This chapter links with Chapter 7 on solving problems in area and volume by providing the foundations of the knowledge needed for that chapter. Functional-skills students will find that the work in this chapter will underpin their work in the applied number Functional skills course that forms part of their Functional-skills studies.

Turning

Turning is something that we take for granted but there are some facts in basic maths that you should understand:

▶ Turns in an anticlockwise direction are classified as positive angles. Turns in a clockwise direction are negative directions.

▶ There are 360° in a complete revolution.

Angles

What is an angle?

An angle is the measure of space that occurs between two straight lines.

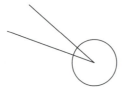

Fig. 15

The diagram shows a complete circle and the angle between the lines.

We split the complete circle or revolution into 360 equal parts, called degrees. So there must be half of 360 degrees on a straight line. Look at this diagram:

Fig. 16

Since there are 360 degrees in a circle, there must be half of this on each side of the line. That means that there are 180 degrees on a straight line.

Notice what we have done here. We began with something we knew and understood, namely that there are 360 degrees in a complete circle or revolution. We then reasoned that there must be 180 degrees on each side of the line that **bisects** the circle. So there must be 180 degrees on a straight line. We have gone from something we knew to something we may not have known. This is fundamental to mathematics and to the style of thinking called mathematical thinking.

Types of angles

Any angle that is less than 90 degrees is called an **acute** angle. An angle of 30 degrees for example is therefore an acute angle. In Figure 16 we have not measured the angle, but we know it is less than 90 degrees, and so it must be an acute angle.

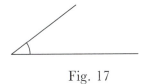

Fig. 17

How did we know that it was an acute angle, without measuring it? Let's imagine a corner, for example the corner of this page. This is a 90 degree angle, known as a **right angle.** The lines of a right angle meet at right angles.

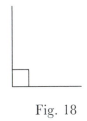

Fig. 18

Incidentally, Figure 18 is also a right angle and not, as one of our students once said, 'a left angle' — there is no such thing!

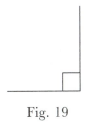

Fig. 19

Angles that are greater than 90 degrees are split into two groups.

Obtuse angles

An obtuse angle is an angle greater than 90 degrees but less than 180 degrees. A typical obtuse angle will look like this:

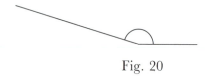

Fig. 20

So, in general terms, an obtuse angle is one that is more than a right-angle (a corner) but less than 180 degrees (a straight line).

Reflex

The third type of angle is called a **reflex** angle. This is an angle greater than 180°. If you are old enough to remember the pop-music of the early 1980s you may remember the group Duran-Duran's hit single, *The Reflex*. This stands out well because it was the one and only time that all our students managed to identify the reflex angle question on the examination paper. They had remembered the name, associating it with something else, in this case a song. If you can associate concepts and ideas with other things, it will help you to remember and to learn.

Finding missing angles on a straight line

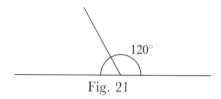

Fig. 21

In the diagram, you can see that one angle is given — 120 degrees — and that there is another missing angle. We already know that there are 180 degrees on a straight line, therefore the missing angle must be 180 degrees minus 120 degrees, in other words 60 degrees.

Again, we are using something that we know to find something that we did not know. This is a fundamental approach in mathematics.

Practice questions
Try these questions.

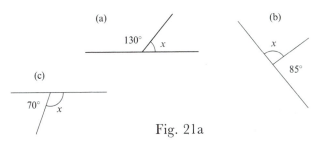

Fig. 21a

Angles around a point
Another very important fact in geometry is that angles around a point always add up to 360°. In Figure 21b, simply add up the angles that you are given and subtract this answer from 360°. Therefore, angle x is 30°.

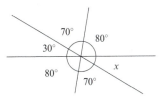

Fig. 21b

Now try these questions

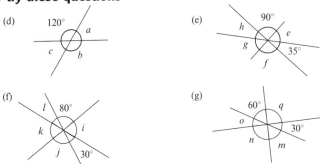

Fig. 21c

Triangles

Names of triangles
You will probably remember some of these names from your school days. Let's remind ourselves of them:

▶ isosceles triangle

▶ equilateral triangle

▶ scalene triangle

▶ right-angled triangle.

Isosceles triangle
This has two equal sides and two equal angles. The equal angles are opposite the equal sides. In the diagram, the two marks on the side show that the sides are equal. This then means that the triangle must be isosceles. It also means that angle *a* and angle *b* are equal.

Remember, in an examination or assessment you will not be told that the triangle is isosceles. You will have to recognise the marks on the sides of the triangle which indicate that they are equal in length.

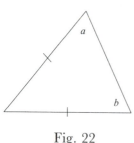

Fig. 22

Equilateral triangle
All three sides, and all three angles, are equal. Therefore, we can easily work out the size of the angle in an equilateral triangle. We said earlier that there are 180° in a triangle. This means that each angle must be 60° (180° ÷ 3).

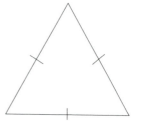

Fig. 23

Scalene triangle
In a scalene triangle, all of the sides and all of its angles are different. Any triangle which has three different sides and three different angles is classed as a scalene triangle.

Right-angled triangle
This is any triangle that contains a right angle.

Solving triangles

Solving a triangle means finding all of the missing sides and angles in the triangle. Finding missing angles in a triangle may seem trivial now that you are engaged in study at a higher level, but as a technique, it is extremely useful.

The first thing to learn is that there are 180° in every triangle. If you want to prove this to yourself, draw a triangle, then rip off the corners. You should find that you are able to line the three corners, the three angles, on a straight line (Figure 24). We already know that there are 180° on a straight line. Since the three angles fit on a straight line, there must be 180° in a triangle.

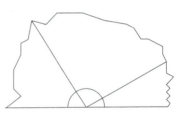

Fig. 24

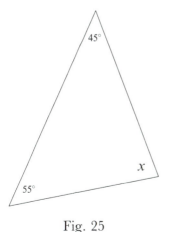

Fig. 25

In Figure 25, two of the three angles are given. Since we know there are 180° in a triangle, we can simply subtract 45 degrees and 55 degrees from 180 degrees. The third and final angle (x) must be 80°. Remember to check that all three angles add up to 180°.

Practice questions

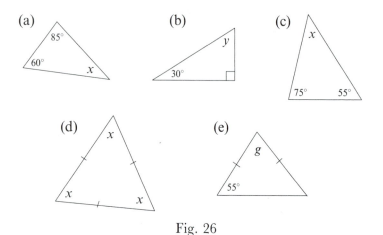

Fig. 26

Parallel lines and angles

Parallel lines are those that remain an equal distance apart from each other for the whole of their length. The classic example is railway lines. Railway lines can curve, but the two rails must remain an equal distance apart, for the train to stay on the track. (In some circumstances, in mathematics, it is possible to show other properties of parallel lines, but this is not appropriate here.)

A line that cuts through two or more parallel lines is called a **transversal**. This line creates some interesting angles.

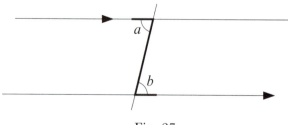

Fig. 27

Look at the diagram carefully. We have added an extra set of lines in the form of a Z. The angles in the crooks of the Z are equal. In mathematical terms, these are called **alternate angles,** but they are also commonly called **Z angles** or the Z-system.

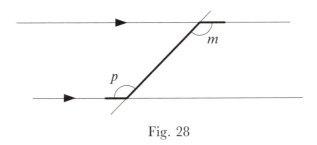

Fig. 28

This diagram shows that sometimes the Z can be backwards. This does not matter. As you can see, the angles in the arms of the Z are still equal.

Corresponding angles

Corresponding angles are also known as F-angles or the F-system. Here, the angles under the arms of the F are equal. In Figure 30, you can see that the F could be backwards, too.

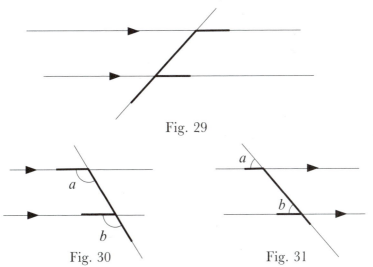

Fig. 29

Fig. 30   Fig. 31

Whenever you are dealing with corresponding angles and parallel lines, always look for the F. Don't be surprised if it appears upside down, or even upside down and backwards.

Now we have covered two types of angles, alternate (Z-angles) and corresponding (F-angles). To make sure you are convinced, please investigate at least six different arrangements of alternate and corresponding angles. In other words, draw six diagrams and then measure the angles in the crooks of the Z and under the arms of the F. This should convince you of the mathematics involved. From the learning point of view, this stage of convincing yourself is important, and you should follow this up by convincing other people. These stages of thinking are essential in developing your mathematical abilities.

Vertically opposite angles

These are angles that meet at a point or **vertex**.

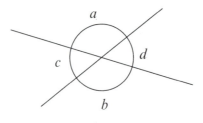

Fig. 32

The diagram shows that the angles that are opposite each other are the angles that are equal.

Allied angles

Allied angles are also known as 'U' angles. Allied angles add up to 180°.

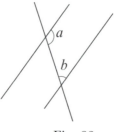

Fig. 33

The diagram shows angles *a* and *b*. Since they are allied angles, the two angles add together to make 180°.

Names of regular polygons

Polygon	Number of sides	Number of angles that are			
		acute?	*obtuse?*	*reflex?*	*right?*
Quadrilateral	4	?	?	?	?
Pentagon	5	?	?	?	?
Hexagon	6	?	?	?	?
Heptagon or Septagon	7	?	?	?	?
Octagon	8	?	?	?	?
Nonagon	9	?	?	?	?
Decagon	10	?	?	?	?

Spend some time investigating these shapes.

▶ Copy and complete the table.

▶ How many of these shapes contain right angles?

▶ Use a computer package to draw regular polygons.

▶ Can you draw a regular polygon with three right angles?

▶ Can you draw a regular polygon with three reflex angles?

▶ Can you draw a regular polygon with four acute angles?

▶ What is the minimum number of sides needed to make a regular polygon?

▶ Try drawing polygons with more than 10 sides.

▶ What is a 12-sided regular polygon called?

Finding the internal angle of a regular polygon

Poly means many, therefore a **polygon** is a many-sided shape. The word **regular**, in mathematics, means that all of the sides of the polygon are the same length. There are many ways of finding the interior or internal angle of a regular polygon. The one we show here (Figure 34) is quite straightforward.

In the diagram, we have an octagon with each of the sides slightly extended. This shows that there are eight external angles. If you travelled around the outside of the octagon, you would travel through 360°. You would also travel through the 8 external angles.

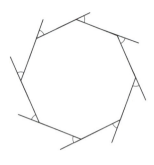

Fig. 34

Therefore, 8 external angles are equivalent to 360°. In mathematics, we use the sign ⇔ to mean equivalent.

$$8 \text{ external angles} \Leftrightarrow 360°$$
$$\text{so } 1 \text{ external angle} \Leftrightarrow \frac{360°}{8}$$
$$\Leftrightarrow 45°$$

But, we want the internal angle. Since the internal and external angles lie on a straight line, and we know that there are 180° on a straight line, the internal angle must be:

$$180° - 45° = 135°$$

Measure the internal angles in the diagram to confirm this is so.

This method works for any regular polygon: simply divide 360° by the number of sides. This gives you the size of the exterior angle. Then subtract the exterior angle from 180°. This will give you the size of the interior angle.

▶ Try this for pentagons: what is the size of the internal angle of a regular pentagon? What about hexagons?

▶ Try finding a regular shape with an internal angle of 144°. How many sides does it have?

▶ What is the largest internal angle that a regular shape can have? How many sides will this shape have?

Investigating different aspects of mathematics is an important part of your learning. We have included these questions to stimulate your thinking and to help you in your learning.

The theorem of Pythagoras

There is some doubt over whether there actually was a man called Pythagoras. Some historians believe that he was a myth created at the time, to protect the identity of others. Others say that he definitely did live and travelled far and wide. It is certainly true that the theorem or rule that we attribute to Pythagoras was well known in Asia and China long before it came to Greece and the West.

Pythagoras' theorem is concerned with right-angled triangles:

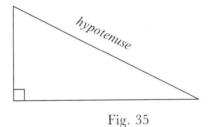

Fig. 35

It is important to notice two basic things about right-angled triangles:

▶ The hypotenuse is the longest side.

▶ The hypotenuse is always opposite the right-angle.

Pythagoras tells us that: 'the square on the hypotenuse of a right-angled triangle is equal to the sum of the squares on the other two sides'.

Finding a missing hypotenuse of a right-angled triangle

Let's look at an example, to explain this:

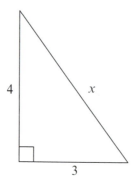

Fig. 36

In the diagram, we want to work out the length of the hypotenuse, marked x. Using Pythagoras:

$x^2 = 4^2 + 3^2$ (remember, 4^2 means $4 \times 4 = 16$)
$x^2 = 16 + 9$
$x^2 = 25$

But we do not want x^2, we want x.

Reminder
When you were at school, you may remember using this symbol: $\sqrt{}$. It is called the square root symbol. The square root of a number is another number which, when multiplied by itself, takes you back to the first number. For example, the square root of 9 is 3, because $3 \times 3 = 9$. Mathematically we write this as $\sqrt{9} = 3$

Back to our solution of the length of the hypotenuse... We had

$$x^2 = 25$$

We do not want x^2, we want x, since x is the square root of x^2. We need to find the square root of 25:

$$x = \sqrt{25}$$

$$= 5$$

So we have now found that the hypotenuse is 5 units long. Note: if we were working in centimetres it would be 5 centimetres long. In inches it would be 5 inches long. This is the famous **3, 4, 5 triangle**. It is used in engineering, and in construction, to guarantee a right-angle in any manufacturing process. It is a very important practical thing to know.

Let's look at another example:

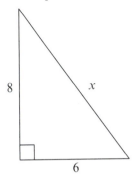

Fig. 37

Use the first example as a model to work this one out. Again, we want the length of the hypotenuse. Have a go at it first, before looking at the solution. If you haven't had a go at this calculation, please try it now, before you go any further.

In this calculation, we want to find the length of the hypotenuse, labelled x. Using Pythagoras:

$$x^2 = 8^2 + 6^2$$
$$x^2 = 64 + 36$$
$$x^2 = 100$$
$$x = \sqrt{100}$$
$$x = 10$$

Really this should not be a surprise, since this triangle is just a bigger version of a 3, 4, 5 triangle. Compare the sides of this triangle with the 3, 4, 5 triangle. What do you notice?

You should have noticed that each of the sides on the second triangle is twice the length of each of the sides on the first triangle. In fact, it is exactly the same triangle, only twice the size. Bearing that in mind, look at this:

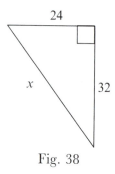

Fig. 38

Without doing any calculation, find the length of the hypotenuse, giving reasons for your decision.

It may look a bit daunting but think about what you know. You know what a 3, 4, 5 triangle looks like. In a case like this, you can bet your boots that this is an enlargement of a 3, 4, 5 triangle. The problem is deciding how *big* an enlargement. Compare the sides, 24 is *eight* times bigger than 3 and 32 is *eight* times bigger than 4, so the hypotenuse must be *eight* times bigger than 5. The hypotenuse is therefore 40.

Finding a missing side that is not the hypotenuse
This question sometimes causes students to panic, but it needn't. Let's take another look at the 3, 4, 5 triangle:

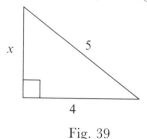

Fig. 39

Here we know that the answer will be 3. The point of this example is to show you how to work out the answer.

Using Pythagoras

$$5^2 = x^2 + 4^2$$

Here the x is on the wrong side of the equation. The problem is that we have the x^2 and the 4 to deal with. They are both on the right hand side of the equation, when we actually just want the x to work out its value.

What we need to do is to subtract 4 from both sides. Subtracting 4 from the right hand side means we will be left with the x^2 and everything else will be on the other side of the equation.

Remember, whatever you do to one side of the equation, you must always do exactly the same to the other side of the equation. So:

$$5^2 - 4^2 = x^2$$
$$\text{or } x^2 = 25 - 16$$
$$x^2 = 9$$
$$x = \sqrt{9}$$
$$x = 3$$

To sum up:

▶ If we need to find the hypotenuse, simply use Pythagoras. Square the other two sides, add them, and find the square root.

▶ If we need to find one of the other sides, again use Pythagoras. Square both sides and then subtract the smaller side from the hypotenuse. But do not forget to square-root your answer.

Tutorial

Progress questions

1. What is the name for the group of angles that are less than 90°?

2. Find the missing angles in these diagrams

(a)   **(b)**

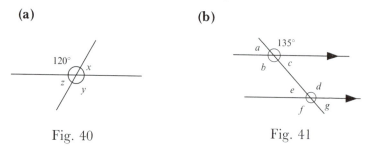

Fig. 40   Fig. 41

Seminar discussion

1. Read G. C. Joseph's *The Crest of the Peacock* published by I. B. Tauris. Discover and be ready to talk about the non-European roots of mathematics.

2. 'We believe that raising the level of mathematics ability in the general population is essential for national defense (*sic*). An uneducated population is the greatest threat to democracy...' Myron Tribus and J. H. Hollomon, p. 19, *Productivity... Who is responsible for improving it?* British Deming Society. Do you agree? If so, why? And if not, why not?

Practical assignments

1. What is the length of the diagonal of the Cardiff Millennium Stadium?

2. Calculate the diagonal length of the front cover of this book.

Study and revision tips

1. Make sure that you practise using the techniques outlined in this chapter.

2. Be ready to use the problem-solving techniques outlined in this chapter elsewhere in this book, and within your course.

3. The assignment on the Millennium Stadium is an example of the application of knowledge in problem solving.

7

Area and Volume

One-minute overview — Area and volume are fundamental in mathematics, but surprisingly many adults find it very difficult to work with these two topics. Area and volume are vitally important in all areas of work. For instance, business premises involve regulations about the floor area per office worker. You may need to work out the volume of storage space in industrial premises or for transport. Suppose you have to transport units of production that each have a volume of $50m^3$, what is the minimum-volume vehicle that you will need to use? If you get it wrong, you could be using a vehicle that is too large, and therefore travels partly empty. This could cost your company or business money. If the vehicle is too small, the operatives will be unable to load everything on board. Understanding area and volume is a valuable skill to have, both in the workplace, and for your course of study.

In this chapter, you will learn:

▶ the definition of **perimeter**

▶ how to calculate areas of shapes

▶ how to calculate volumes of solids

▶ how to find the circumference and area of a circle.

Students calculating in areas and volumes are reminded that:
1. Areas are found in square units, for example cm^2 and m^2.
2. Volumes are found in cube units, for example cm^3 and m^3.

Perimeter

The perimeter of the shape is the total distance round the boundary of the shape. When the dimensions of a shape are known, it is easy enough to calculate its perimeter.

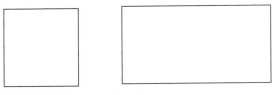

Fig. 42

1. Generally, you find the perimeter of a square by multiplying the length of one side by 4.

2. You find the perimeter of a rectangle by calculating (2 × width) + (2 × length).

Units of area

When dealing with area, we need to be able to convey to other people how big that area is. For this we need standardised measures, clearly defined. A square is a good standard with which to measure an area. Large areas are best found by using large squares while smaller areas are more easily dealt with by using smaller squares. We will need a variety of standard squares to measure areas of all types. The size of the square that would be useful in working out the area of this page is one with a side of 1 cm.

Fig. 43

The area of the square in Figure 43 is: $1 \times 1 = 1$ square centimetre, or 1 cm^2.

Since 1 cm is 10 mm, we know that a square of side 1 cm can be divided into 10×10 squares, in other words 100 squares of side 1 mm each. Each of these smaller squares has an area of 1 square millimetre, or 1 mm^2, so

$$1 \text{ cm}^2 = 100 \text{ mm}^2$$

Fig. 44

Since 100 cm = 1m, a square of side 1m can be sub-divided into 100 x 100 squares. Each of these smaller squares will have a side length of 1 cm and an area of 1 cm^2. So:

$$1 \text{ m}^2 = 100 \times 100 = 10\ 000 \text{ cm}^2$$

The units you use in calculating depend on what you are calculating. For instance, it would be ridiculous to calculate the area of a country in square centimetres. Similarly, one would not try and express the area of this page in square kilometres.

Another measurement to know is the **hectare**. Farm land is now measured in hectares, where

$$1 \text{ hectare (ha)} = 10\ 000\text{m}^2$$

Finding areas

The area of a square

The square in Figure 45 has sides of length 4cm. It is made up of 16 squares, each of side length 1cm.

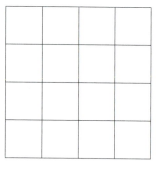

Fig. 45

We can find the area of the big square by multiplying 4cm by 4cm, and reaching the answer of 16cm^2. In general, the area of a square is calculated like this:

$$\text{area of square} = (\text{side length})^2$$

Using algebra, we would write it as:

$$A = l^2$$

where A is the area of the square, and l is the length of one side.

The area of a rectangle

Imagine a rectangle of length 6m and width 3m. It can be divided into 18 squares whose sides measure 1m, so its area is 18m^2.

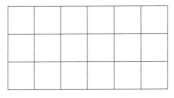

Fig. 46

In general, the area of a rectangle is given by this formula:

$$\text{area} = \text{length} \times \text{width}$$

In algebra, we write it as

$$A = lw$$

where A is the area of the rectangle, l is the length of one side, and w is the width of the rectangle.

Practice questions on area

1. A photographic mount is rectangular and measures 18cm by 22cm. A hole, which is also rectangular, is cut in the mount. The hole measures 9.5cm by 13.4cm. What is the area of card left around the hole?

2. An L -shaped room is made up of two rectangles, one 14m by 8m, the other 16m by 5m. What is the total area of floor space?

3. A square room, of wall length 22m, has a rectangular alcove measuring 2m by 50cm, in addition to the room area. What is the total floor area of the room and the alcove?

4. A rectangular wall has a window in it, measuring 2m by 1m. The area of the window and the wall is 20m². If the room is 5m long:

   (a) How high is the wall?
   (b) What is the area of the wall, without the window?

Finding the circumference and the area of circles

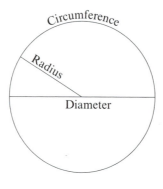

Fig. 47

The diagram shows a circle, with its main parts named. They are:

1. **circumference** – the outer perimeter of the circle and is a continuous length.

2. **diameter** – the distance across the circle from one point on the circumference to another point on the circumference, via a straight line through the centre of the circle. We have shown the diameter going horizontally across the circle. In fact it could go any number of ways. A line going vertically across the circle through the centre point is also a diameter.

3. **radius** – this is the distance from the centre point to any point on the circumference.

Circles and π

When you were at school, you may have done an experiment where you had to place string around circular objects. This gave you a rough measure of the circumference of the object. You then used string to measure off the diameter and find how many times the diameter divided into the circumference.

The idea was to introduce the idea of pi, the number that is given the Greek letter π. If you take any circle, and then divide the circumference of the circle by its diameter, you always get the same answer, namely 3.141592654 (to 9 decimal places). Try it for yourself. π is a number that has never been finally determined, even though it has been worked out to millions of decimal places.

We can use pi to work out the circumference of a circle. There are two formulae for finding the circumference:

$$C = \pi D \quad \text{and} \quad C = 2\pi r$$

where C is the circumference of the circle, D is the diameter and r is the radius.

Generally, we use the first formula when we know the diameter of the circle. For example, find the circumference of a circle of diameter 5cm:

$$\begin{aligned} C &= \pi D \\ &= 5 \times 3.141592654 \\ &= 15.71\,\text{cm} \ (2\ \text{d.p.}) \end{aligned}$$

The diameter is split into two equal radii (singular is radius, plural is radii), so if we are given the radius of the circle, we can use the second formula:

$$C = 2\pi r$$

Again, find the circumference of a circle of radius 4cm:

$$\begin{aligned} C &= 2\pi r \\ &= 2 \times \pi \times 4 \end{aligned}$$

$$= 25.13\text{cm} \, (\text{to 2 d.p.})$$

Since the main theme of this chapter is area and volume, we need to be clear on how to find the area of a circle. Let's do that next.

Finding the area of a circle

There is a separate formula for finding the area of a circle, it is:

$$A = \pi r^2$$

where A is the area of the circle, and r is the radius of the circle. Notice that

$$A = \pi r^2 \quad \text{means} \quad A = \pi \times r \times r$$

Example
Find the area of a circle of radius 5cm. Use the formula $A = \pi r^2$:

$$A = \pi \times 5 \times 5$$
$$= 78.54\text{cm}^2$$

Practice questions

1. Find the area of a circle of radius 6cm.

2. Find the circumference of a circle of diameter 10cm.

3. Find the circumference of a tin of beans, of radius 6cm.

4. Find the area of a circular pond, of diameter 10m.

5. Find the area of a roundabout, of radius 12m.

Cuboids and cubes

Students sometimes have difficulty naming solids. Often, any three-dimensional box shape is called a cube. This is incorrect. For example, a solid with opposite faces made up of parallel rectangles is called a **cuboid**.

Some of the faces of a cuboid may be square. Think about this for a moment. It must mean that *all* of the lengths of the sides of *some* of the faces are equal.

When *all* of the lengths of *all* of the sides are equal, the sides must be made of squares. In that case the solid is a **cube**.

Drawing solids

When drawing solids in mathematics, there are conventions to be followed. Lines that we can see are drawn as solid lines. Lines that are hidden from our view are drawn as broken lines, to show their position.

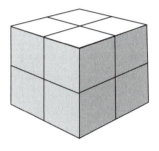

Fig. 48

This diagram shows a cube. However, lines for the edges out of view are not shown. This makes it difficult to view the shape.

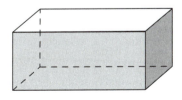

Fig. 49

In this diagram, we can see the hidden edges and so can better judge the shape.

Finding volumes

The volume of a cuboid

Fig. 50

Where all the edges of a cube measure 1cm, the volume of the cube is one cubic centimetre. This can be written mathematically as $1cm^3$.

A cuboid can be built up from cubes of volume $1cm^3$ like this:

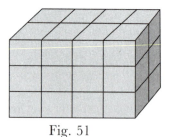

Fig. 51

This cuboid is built up of 24 cubes. In general, the volume of a cuboid is calculated by:

$$volume = (length \times width \times height)$$

The answer is given in 'cubic units'.

Units of volume

The common units in which we measure volume are all based on units of length, for example millimetres (mm), centimetres (cm), feet, and yards. Think of a cube whose edges measure 1cm. Imagine that we can split that cube into 10 layers. Each layer will contain 10 $\times$ 10 cubes with edges 1mm long, or 100 cubes. Therefore $1cm^2$ has a volume of $10 \times 10 \times 10mm^3$. In the same way,

$$1m^3 = 100 \times 100 \times 100cm^3$$

and so on.

Capacity

Students often confuse capacity and volume, when dealing with liquids. **Volume** is the amount of liquid in a container. **Capacity** is the volume of the container. Suppose the petrol tank of your car can hold 60 litres, and that it is half full. Then the volume of petrol in the tank is 30 litres. However, the capacity of the tank – the amount that it is capable of holding – is 60 litres.

Millilitres and centimetres cubed

A millilitre is 1000th of a litre. The 'milli' prefix means '1000th of'. A medicine spoon is usually calibrated to give 5ml. This means that there is a relationship between capacity and units of volume. $1000cm^3 = 1$ litre. This means that $1ml = 1cm^3$.

Surface areas and volumes of prisms

The cross-section of a shape

Imagine a cylinder that is cut across the middle, and then viewed from the cut end. If you viewed the cut end, you would see a circle. This circle is the **cross-section** of the cylinder.

Uniform shapes like cylinders have a cross-section that is the same, regardless of where the cut is made along its length. Other shapes may have a cross-section that varies where the cut is made. The whole-body scanning technique used in medicine, for example, takes cross-sections of the human body. Clearly, these are of different sizes, depending where they are taken along the body length. Shapes which have a cross-section that is the same, along the whole of their length, are said to have a **uniform cross-section**.

Defining shapes with uniform cross-section

Shapes with a uniform cross-section are called **prisms**. For example, a cylinder is a prism. The cross-section of the cylinder is uniform along the whole of its length. Regardless of where the cylinder is cut, the cross-section will always remain the same. In Figure 53, the cuboid is cut at a number of places. Each time the cross-section is uniform. Therefore this too is a prism, a cuboidal prism.

Figure 54 shows the packaging shape of a well-known bar of chocolate. Again, the solid can be cut at various places, and the cross-section will always be uniform. This is a triangular prism. Notice that it is the packaging that is a prism – not the chocolate bar itself. The manufacturers actually make dips in the top of the chocolate bar, thereby ensuring that it does not have a uniform cross-section, so the bar itself is not a prism.

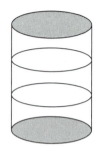

Fig. 52

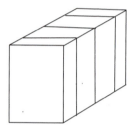

Fig. 53

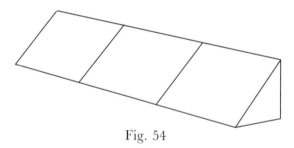

Fig. 54

Finding the surface area of a cylinder

The cross-section of a cylinder is a circle. Finding the surface area of the whole cylinder involves finding the area of the curved surface and then adding it to the area of the two circles that make the top and the bottom of the cylinder.

Since we now know how to find the area of a circle, using the formula

$$A = \pi r^2$$

this is simply a matter of finding the area of the curved surface. Figure 55 shows how the cylinder is made up. The curved surface can be represented as a rectangle. Imagine that the curved surface is unravelled: there are then two circles, from the top and the bottom

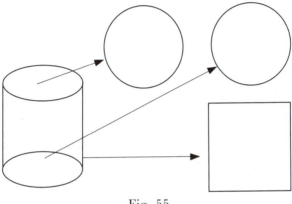

Fig. 55

of the cylinder. The surface area is therefore the area of the rectangle plus the area of the two circles. Algebraically this can be written as

$$A = 2\pi r^2 + \text{the area of the rectangle}$$

The rectangle has one edge which is the circumference of the circle at the top of the cylinder (or at the bottom). Therefore its length must be $2\pi r$. The other dimension is the height of the cylinder, and we can call this h. Therefore, the total surface area of the cylinder is calculated like this:

$$A = 2\pi r^2 + (2\pi r \times h)$$

Finding the volume of a prism

Any prism has a uniform cross-section. Therefore, to find the volume of the prism we use the following formula:

Vol = area of cross-section × length (or height) of the prism

Finding the volume of a cylinder

A cylinder has a cross-section that is a circle, therefore the volume (V) of a cylinder is:

area of cross-section × height

$$V = \pi r^2 \times h$$
$$V = \pi r^2 h$$

For example, find the volume of a cylinder of height 15cm and diameter 10cm. Notice, since the diameter is given, we need the radius.

Therefore, the radius must be 5cm (half of the diameter). Plug this into the formula:

$$V = \pi r^2 h$$
$$= 3.142 \ (\ etc) \times 5 \times 5 \times 15$$
$$= 1178.097245$$

$$= 1178 cm^3 \text{ (to the nearest whole cubic centimetre)}$$

In this example, it is reasonable to round it off in the way we have, because of the context of the question. But we have still shown the full calculator answer, in the second to last line of the calculation. This tells any marker/examiner that a full calculation has been made, and that we have only done the rounding at the end.

What about other prisms?

The principle is just the same. Work out the area of a cross-section and then multiply that by the height. You will then have the volume of the prism.

In the case of the triangular prism (the box for the well-known chocolate bar), the area of the cross-section is the area of a triangle. Since the area of the triangle is ½ × base length × triangle height, then the volume of the prism must be:

½ × base length × triangle height × length of prism.

Try this for a prism of triangle height 5cm, width 6cm and length 10cm:

$$\text{vol} = ½ \times 6 \times 5 \times 10$$
$$= 150 \ cm^3$$

Tutorial

Progress questions

1. Work out the area of the following circles, taking π as the value given on your calculator:

    (a) radius 12cm   (b) diameter 16cm

2. Work out the area of a rectangular room, of dimensions 8m and 4m.

3. Find the volume of a cylinder of height 20cm, and radius 12cm.

Seminar discussion

1. 'The volume of household rubbish that is disposed of into landfill sites could be largely recycled, and used again.' Is this true?

2. The physical volume of hand-held technology, like mobile phones, cannot be reduced indefinitely, because of the size of the human hand needed to operate it. What other functions do you think your mobile will have, as well as phoning, in ten years' time?

Practical assignments

1. Conduct an internet search for information regarding Sir Isaac Newton's work with prisms.

2. Design a package, for a commodity, that is attractive and functional. What is the minimum surface area that is required for your package, to give it the maximum volume?

Study and revision tips

1. When working out questions on area or volume, there is always a temptation to round off answers part way through the calculation. Avoid doing this. It will lead to rounding errors in your final answer.

2. Remember, area is measured in square units, like square centimetres (cm^2) and volume is measured in cubic units, like cubic centimetres (cm^3)

3. If you have rounded off a final answer, always state the degree of accuracy, for example to the nearest whole unit, to 2 decimal places, and so on.

8

Handling Data

One-minute overview — There is a very famous quote of 'lies, damned lies and statistics', which is often wrongly interpreted as meaning that statistics lie. Statistics are simply measures – numbers that indicate pattern, and so cannot lie. Yet liars often use statistics to justify their viewpoint. It is therefore important that you develop an intuitive confidence when dealing with data. This confidence will come from feeling comfortable with data handling because you understand it well. In this chapter we will examine the foundations of data handling. This means developing an understanding of:

► different types of averages, the mean, the median, and the mode

► different types of information known as discrete or continuous data

► displaying data in charts – pie charts, bar charts and histograms

► how to use scatter graphs to show correlation between variables.

Importance of data handling

Data handling is part of every course of study. It is important that you are comfortable with the basic concepts. One word you will come across regularly, and need to understand, is **frequency**, meaning the number of times something happens.

An ability with data handling will also enable you to cope with data used in commerce and in the media. Political information, for example, is often presented in terms of data. Politicians present data in ways that emphasise their own point of view. While main parties may not actually lie with their statistics, they do present their data in a way designed to influence how the public thinks about an issue. Those who do not understand the nature of the data can easily be misled, and if you are misled, it can sometimes cost you dear.

Averages

Types of averages

There are three types of average:

1. mean
2. median
3. mode

But what do they signify? Most people have heard of the word 'average'. Politicians often talk about 'the average man or woman in the street'. But what *is* the average? Let's say that three people earn the following amounts weekly: £300, £120, and £200. To work out the average earnings, add all of the amounts, and then divide by the number of data items:

$$£300 + £120 + £200 = £620$$

Now divide the total £620 by 3:

$$£620 \div 3 = £206.67$$

So, on average each person earns £206.67 (to the nearest 1p). In this case, the average is found by sharing the amounts of money equally between the three people. In reality, none of the people actually earns this amount. The person earning £200 per week comes close to this, but the person earning £300 a week is well above this average. This is only one type of average; in mathematical terms it is called the **mean**.

There are two other types of average to consider: the median and the mode.

1. The median is the *middle* item in a set of data, when the data are placed in order of size, smallest to largest, or largest to smallest.

We will look at some examples to show this later.

2. The mode is the item of data that occurs *most often*. A good example of the use of the mode is in the clothes industry. Clothes manufacturers make their products for the average or modal size man or woman. In the UK, the average size man is about 1.8 metres tall (about 6 feet) whereas the average woman is about 1.65m (about 5 feet 5 inches). Commercially it makes sense to make most clothes to fit people of about this size, since that is the size of most people. High street chains have masses of data on average body size, because it is financially valuable for them to do so. Clearly there are people who are taller or shorter than this, but they are not the majority, so it makes commercial sense to cater for the majority.

Practical importance of the mode

The mode or modal class is important in many ways. The amount of legroom on buses, trains and aircraft is determined by the mode. Indeed, the height of doors is also determined by the modal size of a person's height. You will notice this more if you have ever visited an old property. Often you have to duck through doorways of old houses because, when they were built, the doors were tall enough for the people of the time.

Improvements in diet have meant that the population is now taller than ever and still growing. (Certain 1970s American TV cop shows cheated with their sets. They made the doorways $\frac{6}{7}$ of the real size of a doorway. This made their actors look taller and therefore more macho.)

Politics and TV

At general election times, the phrases 'the average wage' and 'the average person' are used rather freely. As a student, be sceptical when you hear these phrases, because it is not clear what the commentator means by them. Is the speaker referring to the mean of the set of wages, the mean of the set of people? Or are they referring to the mode or the median?

Again, when you hear politicians discussing political matters with journalists, are they using the word 'average' to refer to the same average? A healthy dose of cynicism is important when dealing with and attempting to understand data in the public domain.

Finding the mean

The mean is the arithmetical average, where we add all of the items of data together and divide by the number of items.

Example 1
Find the mean of 1, 2, 3, 4, 5, 6, 7, 8 and 9.

Step one
Add them all together: $1+2+3+4+5+6+7+8+9=45$

Step two
Divide by the number of items: $45 \div 9 = 5$

So the mean of this set of data is 5. The way to understand this is to think of it in terms of real events. Imagine these figures 1 to 9 represented children. If child 1 had 1 sweet, child 2 had 2 sweets and child 3 had 3 sweets, and so on, what number of sweets would each child have, if they were all evened out, and given the same number of sweets? The answer is, they would all have the mean number of sweets – 5 sweets each.

Example 2
Find the mean of the following amounts of money:

£3.20   £35.30   £19.00   £2.50

Add them altogether, $£3.20 + £35.30 + £19.00 + £2.50 = £60$

Now divide £60 by 4 (number of items of data), $60 \div 4 = 15$

So the mean is £15.

Example 3
Find the mean of this set of data: 2 5 7 3 4 9 6 4 5 5

Add them all up: total = 50

Now divide this total by the number of items of data: $50 \div 10 = 5$

So the mean is 5.

Finding the median
Example 1 – finding the median with an odd number of items
Find the median of this set of data: 4 1 5 8 9 7 3 6 2

Remember, the median is the middle term, when all of the items of data are arranged in either ascending or descending order.

Method: Rearrange the items in order: 1 2 3 4 <u>5</u> 6 7 8 9

Therefore the median must be 5, since it is the middle term. This is quite straightforward. All we had to do was put the data in order and then pick the middle item. But what if we are faced with an even set of data? There is no middle item to pick, so we need a slightly different approach.

Example 2 – finding the median in an even set of data
Find the median of this set of data: 1 3 4 2

You still need to put the data in order, either ascending or descending. For example, if we decide to write it in ascending order it becomes 1 2 3 4. But here lies the problem. When we look at the set of data, there is no middle term. Instead, we take the *middle two* terms, 2 and 3.

Find the *mean* of 2 and 3. In other words we add 2 and 3 together, and then divide this answer by 2 (the number of data items)

$$(2 + 3) \div 2 = 2\tfrac{1}{2} \text{ (or 2.5)}$$

So the median of this set of data is 2.5, even though 2.5 was not actually in the original set of data. In short, the median of an even set of data is found by calculating the mean of the middle two terms in that set of data.

Finding the mode

The mode is quite simply the one that occurs the most often and is therefore easy to spot.

Example 1

Find the mode of the following set of data:

1, 2, 3, 5, 7, 1, 8, 1, 9, 1, 5, 1, 76, 1, 89, 1, 234, 1, 66, 55, 7, 33, 48, 39, 1, 2, 55, 1, 15, 12, 1, 4, 1, 6

To work out the answer to a question like this, we draw up a **frequency chart**. Frequency just means the number of times something happens.

Item	Tally	Frequency
1	HHI HHI II	12
2	II	2
3	I	1
4	I	1
5	II	2
6	I	1
7	II	2
8	I	1
9	I	1
12	I	1
15	I	1
33	I	1
39	I	1
47	I	1
48	I	1
55	II	2
66	I	1
76	I	1
89	I	1
234	I	1

From this frequency table, it is easy to see that 1 occurs the most often. Therefore 1 must be the mode.

Example 2
Find the mode of the following:

> blue, blue, red, blue, white, white, green, white, red, white, white, puce, white, pink, white

In this case, it is easy to see that the mode is the colour white. White appears most often.

Finding the range

Range is a measure of the spread of the data. We find it by working out:

> the highest value minus the lowest value.

Note: the 'range' is a single value that we can calculate.

Example 1
Find the range of 1, 2, 3, 4, 5, 6, 7, 8

The highest value here is 8
The smallest value here is 1
$8 - 1 = 7$
So the range is 7

Every year in school examinations, students fail on simple questions like this, because they think of the term 'range' as it is used in everyday life. In mathematics it has a specific meaning.

Practice questions on mean, median, mode and range

1. Find the mean, median, and mode of the following sets of numbers:

   (a) 4, 13, 5, 9, 4
   (b) 7, 22, 3, 17, 3, 13, 7, 11, 3, 4, 9

2. The temperature in °C on 15 days was: 3, 5, 4, 2, 4, 8, 6, 5, 4, 3, 5, 4, 4, 2, 4

   What was the modal temperature?

3. Prakash claims he is better at golf than Andrew. His last five results were 100, 98, 87, 94, 125. Andrew's scores were 101, 110, 98, 97, 98. Is Prakash better than Andrew?

4. The following are the salaries of 5 people working for a small business:

   Fred: £15 500   John: £22 200
   Mair: £24 500   George: £145 000
   Minnie: £12 500

   (a) Find the mean, median, and mode of their salaries.
   (b) Identify which average is not fair and explain in one sentence why it is unfair.

5. Judith is a horticulturalist. She keeps records of her crops to measure their effectiveness. She measured a sample of 25 pea plants and found the following numbers of peas per pod: 5, 6, 5, 4, 6, 7, 8, 6, 5, 4, 6, 5, 7, 4, 6, 7, 4, 5, 6, 7, 8, 5, 4, 6, 4. Find the mean, median and mode of the sample.

Discrete and continuous data

In your course you may have to distinguish between types of data. Data can be grouped into two groups, discrete data or continuous data.

Discrete data

If a set of data is discrete, it can only take certain values. For example, shoe size is discrete. You can have a certain size, or the next size up; there are no continuous measures between these size. For example if you take a size 6 shoe, you can have a size 6, or $6\frac{1}{2}$ or 7. There is no size $6\frac{1}{4}$. The number of students in a lecture hall is also discrete, it must be a whole number of students.

Continuous data

If a set of data is continuous, it comes from measuring, and it can take any value. For example, height is continuous. It is possible to have a height of 1.8 metres and 1.801 metres and so on. The weight of a pear is also continuous, 98 grams or 98.1 grams.

Displaying data

Drawing charts to show data can often be a nightmare for students. One of the problems is that the same chart can have different names in different areas of the country and in different countries. Discuss data handling with your course tutor. Make sure that the names used in your course are the ones that you learn.

An example of the difficulties is what mathematicians and scientists call **exponential curves**. Some social scientists refer to the same curves as **J curves**, because they have an overall J shape.

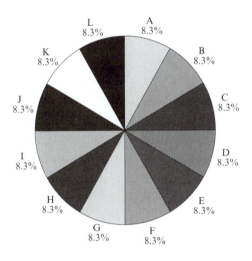

Fig. 56

Drawing pie charts

Figure 56 shows a pie chart. It is so-called because it looks like a pie with slices marked out. You need to be able to draw pie charts by hand but you should also discover how to draw them using a computer package.

Drawing a pie chart

A pie chart is a circle, so it is made up of 360°. This is important to remember, because it affects how we are going to work out the size of each slice.

For example, draw a pie chart for this collection of marks from an access course test:

Marks	0-10	11-20	21-30	31-40	41-50	51-60	61-70	71-80	81-90	91-100
freq	3	5	5	20	25	20	15	10	9	8

Step one
Add up all of the frequencies. You will find that they add to 120 people.

Step two
Remember that 120 people is the whole group, so it is equivalent to 360 degrees.

That means that the whole chart stands for 120 people. Mathematically this is written as:

$$120 \Leftrightarrow 360° \text{ (the } \Leftrightarrow \text{ sign means 'is equivalent to')}$$

We can use this to work out the value of 1 person in terms of degrees. If $120 \Leftrightarrow 360°$, then 1 person must be equal to 3° (360 ÷ 120).

From this we can work out the size of each slice (or sector). Take each of the frequency values in turn. Multiply them by 3, and that will tell you the size of the angle needed at the centre of the circle. Therefore the angles are:

$$9° \quad 15° \quad 15° \quad 60° \quad 75° \quad 60° \quad 45° \quad 30° \quad 27° \quad 24°$$

Now draw a circle. Draw a line from the centre point to the circumference (the boundary of the circle). Use this as a base line. Use the figures just calculated to draw the slices. Your finished pie chart should look like that in Figure 57.

The purpose of the pie chart is to show the statistics visually, to give us an overview of the trend shown by the data.

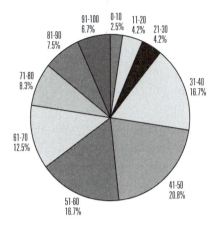

Fig. 57

Bar charts

A bar chart is a simple way to show data. It is usually the first type of chart that children learn to draw when at school. An example of a bar chart is shown below. Notice that the bars on a bar chart are separate. This is because bar charts are used to show **discrete** data (data that is grouped) and not continuous data.

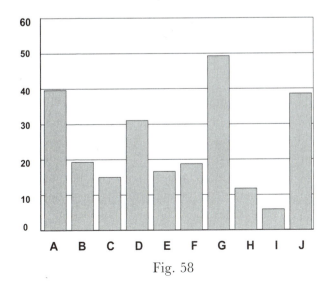

Fig. 58

Bar charts are quite easy to work with. The height of the bar represents the frequency of the event being shown.

Example 1

The table shows how a group of 50 people travel to work. Draw a bar chart to show the data:

Modes of transport	*Car*	*Train*	*Bus*	*Bike*	*Walk*
Frequency	10	12	17	7	4

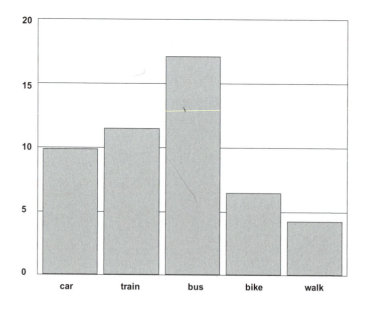

Fig. 59

The bar chart immediately shows the most popular/effective means of travelling to work, visually, because it is the one with the highest bar.

Example 2
In an experiment, two dice were thrown 50 times and the total score was recorded:

2	4	5	7	8	5	8	3	9	7	6	7
5	7	10	7	3	6	5	7	2	7	5	6
12	9	4	11	10	6	8	9	7	7	11	7
5	7	9	10	7	4	2	6	8	2	3	7
8	9										

(a) Draw a tally chart to show this set of data.

(b) Draw a bar chart to show this information

Score	Tally Marks	Total
2	IIII	4
3	III	3
4	III	3
5	HHH I	6
6	HHH	5
7	HHH HHH III	13
8	HHH	5
9	HHH	5
10	III	3
11	II	2
12	I	1

The tally chart sorts the information. Now we are ready to draw the bar chart (Figure 60). Draw the **axes**. The axes are the line across the bottom and the line up the left-hand side. Each on its own is called an **axis**. The word **axes** is the plural.

Example 2
The bar chart in Figure 61 shows the profit/loss made by a bookshop from September 1998 to April 1999.

(a) Estimate the total profit in this period.

(b) Describe what is happening to the shop's profits in this period and provide an explanation for the shape of the bar chart.

Understanding Maths

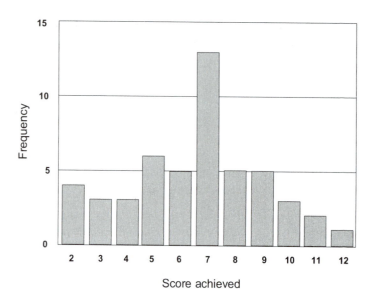

Fig. 60

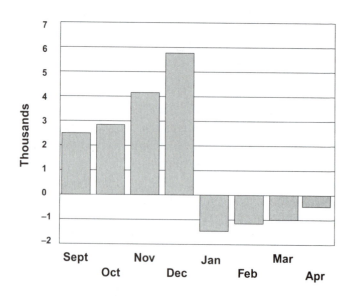

Fig. 61

Grouped data

Sometimes a distribution of a set of data can be spread over a wide range of values. In this type of circumstance, we usually **group** the data.

Example 1

Judith is a gardener and grows courgettes. She tried an experiment on 100 courgette plants. She made up a special fertiliser recipe to see if the plants would grow well, and then weighed all of the courgettes grown with this recipe. As an experimental control she also grew 100 plants without the fertiliser. The results for the first set of plants were:

100	110	125	80	78	98	89	86	98	101
102	105	98	56	87	105	107	101	106	112
80	92	97	96	99	97	85	106	110	101
87	84	78	98	96	93	82	75	79	89
102	106	115	114	120	98	95	91	89	81
85	87	86	78	89	95	101	94	73	108
99	105	104	108	113	112	110	121	107	92
98	78	89	74	75	89	98	96	91	90
87	88	96	94	82	87	81	72	76	72
101	104	107	109	105	88	109	101	103	89

Make a tally chart using the groups given and then draw a frequency chart to show this data.

Notice how the groups are listed in the weights column. The first listing gives weights between 50 and 60 grams. In fact, if you look more closely, they are more closely defined than this, the group is written as, $50 \leqslant w < 60$. This means that this group includes every weight from 50 grams up to but not including 60 grams.

The sign $\leqslant$ means 'less than or equal to'. Therefore $50 \leqslant w$ means that the weight of 50 grams is less than, or equal to, the weight of the courgettes.

The sign $<$ means 'less than', so the weight of 60 grams is not included in this first group.

Weights	Tally	Frequency
$50 \leqslant w < 60$	I	1
$60 \leqslant w < 70$		0
$70 \leqslant w < 80$	HHI HHI IIII	14
$80 \leqslant w < 90$	HHI HHI HHI HHI III	23
$90 \leqslant w < 100$	HHI HHI HHI HHI HHI	25
$100 \leqslant w < 110$	HHI HHI HHI HHI HHI IIII	29
$110 \leqslant w < 120$	HHI I	6
$120 \leqslant w < 130$	II	2
		Total 100

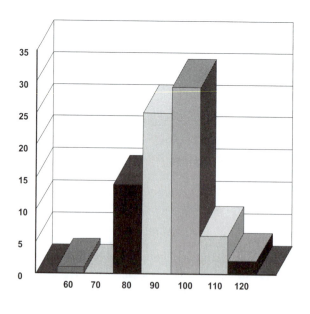

Fig. 62

In this example we have drawn the diagram in three dimensions, simply to make an impact, but you can draw the same chart in 2-D.

Notice that the class intervals are equal. In other words, all of the groups are the same width, across the *x*-axis of the chart. This is important because it is the area under the bar in a histogram that indicates the frequency of the chart. If the width of the group is doubled, then the height of the bar must be halved.

Let's look at one group, to explain this fact. Imagine we repeat the chart above but have all of the classes, except one, to be zero, then the chart will look like this:

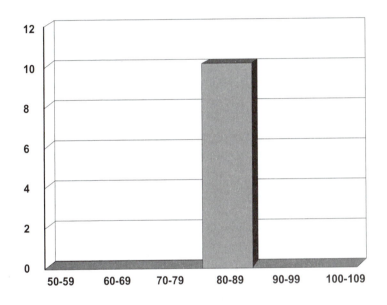

Fig. 63

But if the class interval is doubled, then the height of the bar is halved. This maintains the same area under the bar and so the frequency stays the same. Now the chart looks like the one in Figure 64.

It is quite common, in an exam question like this, to be given data of one trial to draw a chart. You are then given a chart and have to use it to determine which trial it refers to. In this case the second chart would be the **control group**, but we have spared you the hassle.

Variables and correlation

You have probably heard public speakers say things like, 'There is no correlation between low income and crime'. But what does a state-

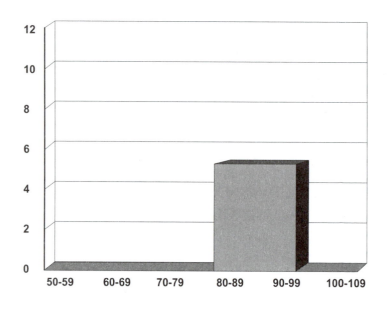

Fig. 64

ment like this mean? What does the word **correlation** mean? It means that there is a connection between two things, and the connection is mutual.

For instance, there is a correlation between people's height and their weight. The taller someone is, the more you would expect them to weigh. Similarly there is a correlation between height and arm span. Taller people usually have longer arms than smaller people and so usually have a wider arm span.

The fact that a correlation exists does not always define the connection. For example, there is a correlation between cigarette smoking and lung cancer. However, no one has actually been able to identify any particular chemical in cigarettes that causes the cancer, yet the figures show that out of every three smokers, one will die from a smoking-related disease and the other two will have serious illnesses. Only a fool will tell you that smoking does not contribute to lung cancer, because they are undoubtedly correlated. Yet there are some poor souls who never smoke a cigarette in their lives, and they still develop lung cancer. There are others who will smoke all of their lives and never develop lung cancer – but it is an observed fact that most people who do smoke are more likely to fall ill.

Odd correlations

There are some odd correlations. For example, it is said that the length of skirts correlates with the state of the economy. Throughout the twentieth century, when women tended to wear skirts below the knee, the economy was in recession. When women wore their skirts shorter, the economy boomed. The so-called swinging sixties is a good example. The mini-skirt was in fashion and the economy was booming. In the early 1970s, the economy went into recession and fashionable skirts became longer. Nobody is suggesting that this is significant, but the correlation does exist.

Showing a correlation between two variables

Variables are items of data that can take any number of different values. For example, the age of people is a variable. If you were to compare age against height, then the variables would be age and height.

To show correlation, we use a chart called a **scattergram** (also known as a **scatter diagram**), shown in Figure 65.

Example 1

Figure 65 shows what we call a **weak positive** correlation. There is a connection between the variables but it is a weak one, since the data is spread out across the chart.

The scattergram in Figure 66 shows a **strong positive** correlation between the variables. It means that there is a strong connection between the variables – as one increases, so does the other, in direct proportion. A good example is height and weight. If a person's height increases through childhood, that person's weight will also increase.

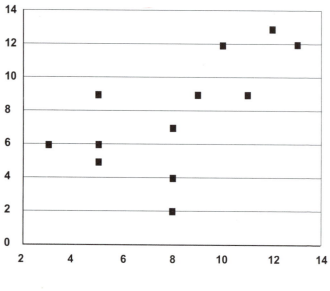

Fig. 65

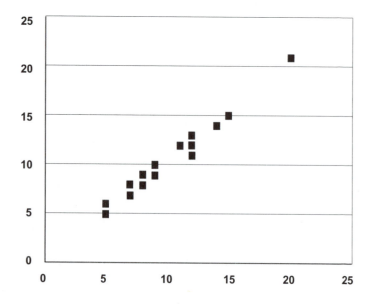

Fig. 66

Figure 67 shows the variables spread unevenly across the chart. This means there is no correlation between the variables.

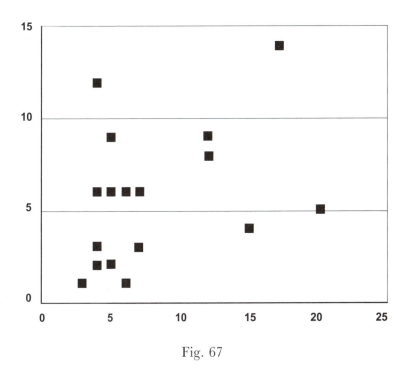

Fig. 67

In the chart in Figure 68 there is a direct correlation between the variables, but it is different from the one noted above. In this case, as one value increases, the other decreases in proportion. This is called a **negative correlation**. A good example of a negative correlation is the one between life expectancy and age. As a person grows older, his or her life expectancy will decrease. So,

▶ A **positive correlation** stretches across a scattergram from bottom left to top right. The tighter the points are on the scattergram, the greater is the positive correlation between the variables.

▶ **No correlation** exists where the points are randomly spread across the chart.

▶ A **negative correlation** exists where the points are spread from top left to bottom right across the chart.

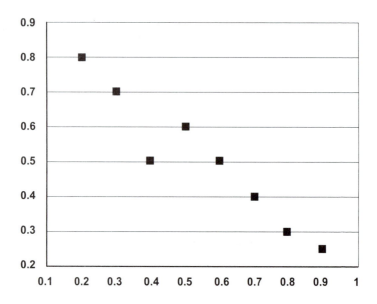

Fig. 68

Tutorial

Progress questions

1. In a survey, the number of occupants in a car was noted. The results were:

> cars with 1 person, 8
> cars with 2 people, 6
> cars with 3 people, 4
> cars with 4 people, 0
> cars with 5 people, 2

(a) Draw a bar chart to show this data.

(b) How many car occupants were surveyed?

2. The marks of 9 students in two papers of an exam were as follows:

Paper 1	25	34	62	68	98	45	78	76	71
Paper 2	27	31	56	60	92	41	83	79	74

    (a) Draw a scattergram to show this data.

    (b) Comment on the nature, if any, of the correlation between these variables.

Seminar discussion

1. When politicians are using statistics to justify their own political views, assuming that they are not actually lying, do you think they are being dishonest?

2. 'Newspapers that follow a political view often distort the statistics of a story.' Is this true? If so, why? If not, why not?

Practical assignments

1. Conduct a survey in the car park of your place of work or college. What is the modal colour of car?

2. Find out the average amount that you and your friends spend on alcohol each week. Which type of average will reflect this statistic the most meaningfully?

Study and revision tips

1. Learn how to use the statistical functions on your calculator. It will save you hours of work.

2. Collect examples of spurious data. You may be able to use it in coursework.

3. Never lie with statistics. We came across a headline in a Midlands newspaper that said that, for the first time since Victorian times, the mortality rate for young men in the area was rising. When we approached the journalist, she referred us to a well-known speaker who had been quoted in her article. When we asked for the data, it could not be found. If you lie with statistics, someone will find you out.

Answers to progress questions (p. 127)

1(a)

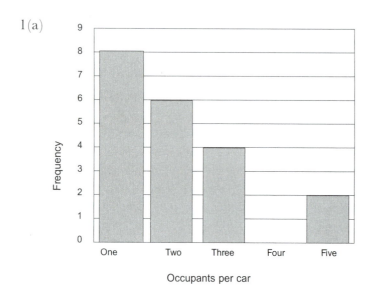

2(a)

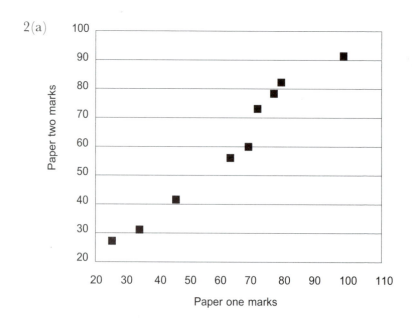

9

Analysing Data

One-minute overview — In Chapter 8, we looked at the different kind of average (mean, median and mode or modal class). Now let's build on that knowledge, and see how to analyse data. Data analysis today is the bedrock of business and industry. Markets are changing so fast that companies need the best possible understanding of the factors of variation within the manufacturing process. Japanese companies have led the way in managing variation; many practise the principles of the statistician Dr W. Edwards Deming, who taught that higher quality products actually meant *lower* manufacturing costs. He realised that the key to better productivity was better management of variations in the production process. Increased productivity means not just more capital, it means better management of systems, staff and machines. A well-known car manufacturer produced a family car with engine cylinders milled within $\frac{1}{16}$ of a mm. The pistons that went up and down these cylinders were similarly milled. If you were unlucky enough to buy a car with a cylinder milled at one extreme and a piston at the other, then your car would burn oil from day one. The management failed to understand the importance of statistical variation. In this chapter, we will study:

▶ measures of central tendency

▶ measures of spread

▶ data-handling techniques.

Measures of central tendency

In Chapter 8, we looked at the mean, the median, and the mode of a set of data. These three measures were introduced as **averages**. We said it is vital to understand each of these concepts, particularly when politicians and so many others refer to the 'average man or woman', or the 'average take home-pay'.

You also need to be familiar with the idea of a **spread** of data. Take, for instance, the spread of marks in an examination, from the lowest to the highest value. In the last chapter, we dealt with one measure of spread, the **range**. This is the highest value minus the lowest value, and is a single value.

Remember:

1. The **mean** is the arithmetical average, where all the data are summed (added up).

2. The **median** is the middle value of a set of data, where the data are arranged in order of size. This order is preferably from lowest to highest, although it can be from the highest to the lowest value. The median is a useful measure when dealing with pay bargaining, for example, because it gives a measure of the middle of the spread of data.

3. The final kind of average is the **mode**. This is the item of data that occurs the most often. The clothes industry, for example, manufactures its products for the 'modal' sized customer, in other words the size that occurs most often.

Each of these measures – the mean, the median, and the mode (not the range) – is a typical mark that neatly summarises the whole set of the data. In statistical terms, they are called 'measures of central tendency'. Each of these measures of central tendency has strengths and weaknesses.

The mean

There is a formula for calculating the mean:

The Greek letter μ (pronounced 'mu') is used to represent the mean.

The Greek letter Σ (pronounced 'sigma') is used to stand for 'the sum of', X is used to indicate the total and N is the number of items of data.

ΣX means the total of all of the items of data, so the mean, μ, is the sum of all of the items of data, divided by the number of items of data. In mathematical terms, this is written as:

$$\mu = \frac{\Sigma X}{N}$$

The **mean** of a set of data can also be shown as $\bar{x}$ and this is a form with which you should be familiar.

Most people who talk of an average are referring to the mean. The weakness of the mean is that it is subject to extreme values; in other words if there is one extreme value in the distribution it will affect the mean value.

For example, let's find the mean salary of the following people: Fred who earns £12 000 per annum, Mary who earns £15 000 and Gareth who earns £325 000. You will find that Gareth's salary distorts the mean as a measure of central tendency of the other two values. The same is true of a very low score. Work this for yourself, to see what we mean.

The mode

The **mode** is also a good measure of central tendency, but again has its weaknesses. For instance, in an examination, if the modal score was zero, this is hardly a good representation of the central value of the scores. This can also be referred to as the modal class.

The median

The **median** is simply the middle term, when the data are in order, but remember you may have an even set of data, in which case you need to find the mean of the middle two terms.

Measures of spread

Another useful measure for summarising a set of data is 'measures of spread' — how scores are spread out in a distribution.

1. A small spread of results is usually seen as a good thing. It

indicates that all the aspects of the distribution, whether people or products, are behaving in a more or less similar way.

2. A large spread could be a problem since it indicates large differences between the individual results in a distribution. This can occur in examples like the one used earlier, where one person's pay is far higher than that of his or her colleagues, resulting in a large spread. In a large spread, the mean is not as representative or meaningful.

The range

Range is the simplest measure of spread. The range is the difference between the highest and the lowest scores, and is a single value. For example, in a distribution where the highest value is 60 and the lowest value 20, the range is 40. It is not '20 to 60'. We stress this because each year, in examinations, many students get this wrong.

As a measure of spread, range tells us the boundaries of the distribution, but it tells us very little else. For instance, we know nothing about the general spread of the results. There could be interesting data in the distribution, but the range only considers the highest and the lowest values, so it is a rather crude, limited measure.

Quartiles

A different way to look at the spread of a set of results is to work out the quartiles. We mentioned the median earlier. You will remember that this is the middle value, when the data are in order of highest to lowest or preferably lowest to highest. The median therefore has the effect of cutting the spread of data into two halves. Quartiles simply cut these halves into quarters.

For example, suppose we have a list of examination results for 100 students, arranged in order of size from lowest to highest. The results were:

0, 9, 10, 10, 12, 12, 14, 14, 15, 16, 17, 18, 19, 21, 21, 21, 22, 24, 24, 25, 26, 28, 31, 32, **36, 38**, 40, 47, 51, 52, 52, 53, 54, 55, 56, 56, 56, 56, 57, 58, 58, 58, 58, 59, 59, 59, 60, 62, 64, **67, 68,** 69, 69, 69, 69, 69, 69, 69, 70, 71, 72, 72, 73, 74, 75, 75, 75, 76, 76, 77, 78, 78, 78,

79, 79, **79**, **80**, 81, 81, 81, 82, 82, 83, 84, 85, 85, 85, 85, 86, 89, 89, 89, 90, 90, 91, 91, 92, 92, 93, 94, 95

From the list above, one quarter along the list lies between the 25th and 26th person's score. So the first quartile is midway between 36 and 38 (we have emboldened the numbers in the distribution, to make it easier for you to see them). The first quartile is therefore 37. Similarly the second quartile is between the 50th and 51st person, because we know it is the median and is therefore half way through our distribution. So the second quartile lies between 67 and 68 and is therefore $67\frac{1}{2}$.

Finally the third quartile lies between the 75th and 76th score, and is therefore $79\frac{1}{2}$.

A word of caution about quartiles

The scores in the examples above were obtained by people. If we are to split the distribution up into quarters, students often ask why the first quartile is between the 25th and 26th value, when $\frac{1}{4}$ of 100 is 25? The answer lies in ensuring there are exactly 25 people in each quartile, in this example. If we made the boundary between the first quartile and the second quartile to be exactly on the 25th person, which quartile would this person's score be in?

Clearly, in a distribution of 100 people, the 25th person does need to be in the first quartile.

Notation

In quartiles, we use the symbol Q to stand for quartile. The first quartile is therefore called Q_1, the second quartile is Q_2 and the third quartile is Q_3. The first quartile can also be called the lower quartile and is abbreviated to LQ. Similarly the third quartile Q_3 (also known as the upper quartile) is abbreviated to UQ.

The interquartile range

This is a slightly more sophisticated measure of spread than the range. It is the range of half of the scores in the distribution, concentrating on those in the middle of the distribution. It is more useful than the range, because it won't be affected by very high or very low values. It will therefore probably give a better picture of the spread of the results. Mathematically, the interquartile range is found by:

$$Q_3 - Q_1 = \text{I.R.}$$

The semi-interquartile range
This is simply half of the interquartile range. It is found mathematically by:

$$\frac{Q_3 - Q_1}{2} = \text{S.I.R}$$

Why use quartiles?
There are a number of benefits:

(a) Quartiles can tell us about the distribution, and if the distribution is symmetric about the median within the interquartile range.

(b) We can use the difference between Q_2 and Q_1 to tell us the range of one quarter of the scores below the median, and the difference between Q_3 and Q_2 tells us the range of one quarter of the scores above the median.

Variation
What we really need is a measure of spread that takes into account each and every score. By starting at the mean for the distribution, we can work out the deviation away from the mean by simply calculating $X - \mu$ where X is the score and μ is the mean. Although this can be done for every score, when they are all added, they tend to cancel each other out.

For instance, in a distribution with a mean of 30, a mark of 55 will give a deviation of 25, but a mark of 5 will give a deviation of –25. This means that when the scores are summed, they will cancel each other out.

This is undesirable, since it does not show the variability within the distribution, as it really is. Since the mean is the 'balance' position of all of the results, adding up all of the deviations will result in a total of zero, because all the positive deviations will be cancelled out by the negative deviations. If you are not convinced by this, try some yourself.

For example, take the distribution:

$$2, 4, 6, 8$$

Find the mean of this small set of data, by adding them up and dividing by 4. You should have found that they sum to 20, and therefore the mean is 5. Now work out the deviations from the mean:

Score	Deviation $\Sigma(X - \mu)$
2	-3
4	-1
6	1
8	3
$\Sigma(X - \mu) = 0$	

This is hardly useful, is it?

If you look at this simple distribution, you will see the negative deviations from the mean.

The point is, a negative deviation tells us how far a score is below the mean. In fact, what we really want to know is how far away from the mean a score is, and not really whether it is above or below the mean. We need to be able to add the deviations so that they do not cancel each other out, to give a reasonable estimate of the real variability of the scores.

There are two ways forward: **absolute deviation** and **variance**.

Absolute deviation

In this method, we ignore the negative sign and treat all values as positive. There are two vertical lines placed either side of the calculation, like this $|X-\mu|$ This gives us a measure of what is called the absolute deviation.

If the absolute deviation is divided by the number of scores, the result is the mean absolute deviation. Mathematically this is written as:

$$\text{Mean absolute deviation} = \frac{\Sigma |X - \mu|}{N}$$

where Σ means 'the sum of', in this case the sum of the positive deviations, and N is the number of items in the data.

Variance

An alternative to the 'mean absolute deviation' is to take all of the deviations and then to square them. A negative deviation squared will give you a positive answer, so the sum of the squares of the deviation will also give you a positive answer. The way to do this is:

1. find each deviation from the mean

2. square each deviation

3. add up the squared deviations.

If we then divide this figure by the number of items of data, we arrive at a figure which is the mean of the squared deviations. This mean of the squared deviations is called the variance. Mathematically this is written as:

$$\text{Var} = \frac{\Sigma (X - \mu)^2}{N}$$

The variance tells us the average variability of the scores around the mean, when they are expressed as squared deviations. The variance is also a good measure of spread, because the variance is large when the data are spread out far, and small when the data are closer.

Standard deviation

The variance is a measure of the squared deviations from the mean, so it cannot tell us the distance from the mean in a frequency distribution. In order to find the distance from the mean, we need to find the square root of the variance, and this is called the standard deviation. Mathematically this is written as:

$$\sigma = \sqrt{\frac{\Sigma (X - \mu)^2}{N}}$$

where σ, the lower case Greek letter for sigma, is the standard deviation.

Let's look at a straightforward example. Here, if we sum the scores, they come to 15. Since there are 5 sets of data, the mean must be:

3                    $\left(\frac{15}{5} = 3\right)$

Score	Deviation	Squared deviation $\Sigma(X - \mu)^2$
1	-2	4
2	-1	1
3	0	0
4	1	1
5	2	4
		$\Sigma(X - \mu)^2 = 10$

Now, dividing the sum of the squares, 10, by the number of scores, 5, gives a variance of 2. Taking the square root of the variance gives a standard deviation of 1.41.

The standard deviation gives a measure of spread about the mean. Generally, in most cases, most of the results will lie within ± 1 standard deviation of the mean. The standard deviation gives a distance that is standard from the mean, for that set of data.

There is a potential problem in dividing by N, of which you must be aware. You should only divide by N when:

(a) the values in the distribution concern the whole of the population

(b) the values in the distribution represent a sample from the population, and we are only investigating the variation and the standard deviation within the population itself.

When investigating the variation and standard deviation with the population, by taking a sample of the population, you should always divide by $n-1$, where n is the size of the population. The reasons for this are quite complex, and to do with the required degrees of freedom. A standard deviation that is calculated on such a sample, drawn from the larger population, is called the 'sample standard deviation'.

Using statistics in management

All processes in all industries show variation. Whether you manufacture left-handed sprockets for widgets, or you work in a service industry, daily life always throws up variations in performance. The skill of management is to manage those variations and keep them within acceptable limits.

To manage properly, therefore, calls for an understanding of variation and, in turn, of statistics. Modern managers are developing a better understanding of 'statistical process control', to control and reduce the variability in their processes.

Numbers and data

Numbers and data are not the same thing. **Data** are almost any kind of information useful in solving problems — and of course it often includes numbers. Numbers alone are often meaningless measures, or counts, which can confuse. Data arising from counts are called **discrete** data and can only occur at definite points. For example, there can be 0, 1, 2, 3, 4 etc mistakes on a page. There cannot be 3.768 mistakes on a page.

Data that result from measurements on a continuous scale are called **continuous data** or **variable data**. Some scales are measured in discrete quantities so it is inappropriate to define all continuous data as being measured on a scale; the scale must be continuous. For example, British shoe sizes are measured on a scale, but the scale is discrete. For instance, you may take a size 6, or a $6\frac{1}{2}$, but you cannot take a size $6\frac{3}{4}$ since such a size does not exist in the shoe industry.

Data-handling techniques

There are some basic data-handling techniques that many people at work need to understand and practise. They include:

▶ process flow charting

▶ tally sheet monitoring

▶ histograms

▶ Pareto analysis

▶ cause-and-effect analysis, using Ishikawa diagrams

▶ scatter diagrams

▶ control charts

Process flow-charting

Flow-charting is a technique for depicting the steps or elements involved in a process of some kind. For example, at school you may have drawn a simple flow chart to show the steps involved in making a cup of tea. In industry, you will need to draw flow charts to describe processes in the workplace. The symbols used in a flow chart are shown in Figure 55.

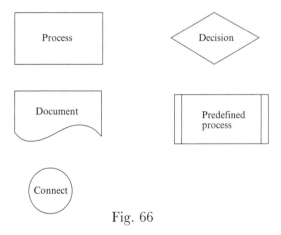

Fig. 66

Good flow-charting skills are essential for managers wanting to systematically plan or examine any process.

Many people find, when they sit down to write a flow chart of a system they use every day, that:

1. they do not totally understand the system themselves

2. they are unable to complete the flow-chart without the help of others.

Drawing up flow charts is very good management practice. The charts may well reveal the needless movement of goods and materials, and trigger good suggestions for cutting out waste. Each stage in a flow chart offers managers a chance to check the quality and efficiency of the process.

Each stage should see the next stage in the flow as its 'customer'. Each stage then needs to develop a set of success criteria that have measurable outcomes, against which the quality of the process can be measured. This is taking us into the blurred edges between mathematics and management mathematics, but we wanted you to be aware of the application of mathematics in this form.

Tally charts

Tally charts are a tool for data collection, and they are probably the most logical place to start in problem-solving or process control. It is often the case that personal assumption about a defect in a process is not borne out by the facts. Tally charts are developed by:

(a) making a firm agreed decision about the event to be monitored

(b) making a firm decision about the frequency of data collection and the time period over which the data will be collected

(c) using a straightforward form on which to collect the data

(d) collectors having integrity to collect the data honestly, and record it accurately

(e) following up the recording of the data by analysing the data.

Histograms

These were discussed in Chapter 8.

Pareto analysis

The Pareto principle states that 80% of results tend to come from 20% of efforts. In Pareto analysis we aim to identify the symptoms or causes of defects in output. Generally this results in identifying that the bulk of the defects are caused by a few of the causes. It then becomes easier to identify and remedy the causes of the problems.

Ishikawa diagrams

Ishikawa (or fish-bone) diagrams are used to illustrate the inputs which affect the quality of a process. See Figure 56. It is good practice to work through an Ishikawa diagram and to isolate each of the variables. Each of the sub-arrows are the sub-causes of principal causes. These can then be extended to sub-sub- causes and so on.

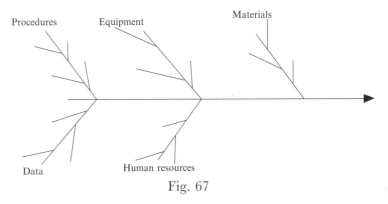

Fig. 67

Scatter diagrams

These are also known as scattergrams and show correlation between two variables. This was discussed in Chapter 8.

Control charts

Control charts are designed to track the variability within a process. They use small samples taken at random. The effect of the control is to keep the process within acceptable boundaries of error. Unfortunately, an isolated result may make you think that a process is going out of control, and lead you to halt it, when in fact there is nothing wrong with the process.

QCAI

All of these stages form part of the process in statistics which can be referred to as QCAI. This is a thinking model, meaning Query, Collect, Analyse and Interpret.

Query

This is where a hypothesis is made, or question asked. It is vital to be clear what the question or the hypothesis actually is. Everyone involved in the research should agree with using the query as the starting point.

Collect

We collect data, for example by using questionnaires or tally charts.

Analyse data

We analyse the data, and represent our findings in charts and graphs.

Interpret data

Interpreting data means explaining what the patterns revealed in the graphs and charts actually mean.

Tutorial

Progress questions

1. Explain the positive and negative benefits of using the mean, median and mode as measures of central tendency in a set of data.

2. Why should we use the sum of the squares of the deviations from the mean, in a set of data, rather than the deviations themselves?

3. Explain what difference occurs in the use of the formulae, when calculating sample variance and sample standard deviations.

Seminar discussion

1. 'Variation in a process is manageable, by using basic statistics.' True or false?

2. 'The mean of a set of data is more information than most people need.' Discuss.

Practical assignments

1. Draw a flow chart to illustrate a process used in your study area or course.

2. Draw a flow chart to illustrate the process of getting ready and leaving your home, to arriving at college or university.

Study tips

Apply the techniques described in this chapter to finding the mean, median and mode, and variance and standard deviation within the context of data generated within your course, or area of study. Make sure you gain the experience of using these techniques in context.

Algebra

One-minute overview — You may well have sighed with relief when your school maths was over, and are now questioning the need for a chapter on algebra in this book. The aim is to refresh your basic knowledge. Although you may not have to solve equations in your course, you will certainly have to handle data. In this, you may well have to use spreadsheets, where you will have to write and rearrange formulae. Using a spreadsheet is an essential skill in business — vital in managing cashflow, for example. Cash flow is the planned expenditure and income of a business compared with its actual expenditure and income. The aim is to keep the business solvent and within budget. Businesses that fail usually do so because their cash flow dries up. Too much money has to be paid out before sufficient money comes in. Certain businesses require heavy expenditure at the start, and have to keep a very close eye on where the money is going. One failed businessman said that he found 'cash flow is more like cash ebb'. The aim of basic algebra is to give you the background to manipulate and rearrange formulae, so you can use these skills elsewhere, especially with spreadsheets. In this chapter we will look at:

▶ multiples and factors

▶ solving different types of linear equations

▶ changing the subject.

Multiples and factors

A **multiple** of a number is the answer in the times table of that number. For example:

▶ the multiples of 5 are 5, 10, 15, 20 and so on

▶ the multiples of 2 are 2, 4, 6, 8, 10 and so on.

A **factor** is a number that goes exactly into another number. For instance, the factors of 6 are 1, 2, 3, 6. These are all numbers that go exactly into 6, leaving no **remainder**.

Solving linear equations

Solving equations with x on one side
An equation is like a balance:

> *Key point* – Whatever you do to one side, you must also do to the other side.

Older readers will remember the old fashioned greengrocer scales. The greengrocer would place the produce on one side of the scale and balance them with iron or brass weights on the other side of the scale. Younger readers may find that the domestic scales in their family kitchen work in the same way. This is a good way to think about solving equations.

Example 1
Let's look at this example:

$$x + 4 = 10$$

In plain English, it means: take an unknown number (x), add 4 to it, and the answer you get is 10.

Clearly the missing number has to be 6, since $6 + 4 = 10$. But bear with us, because we need to make clear the method of solution. We are using an easy question because we want you to concentrate on the method — you can tell by looking that the x must be 6. So, starting again, we have:

$$x + 4 = 10$$

Now, look at the left-hand side. It says $x + 4$, but we only want to know about x.

Let's take away 4 from the left hand side. Since 4 - 4 = 0, that will leave the x on its own on the left-hand side.

> *Remember* – Whatever you do to one side of an equation, you must do exactly the same to the other side.

So we must also take 4 from the right-hand side. The next line becomes:

$$x + 4 - 4 = 10 - 4$$

Remember, on the left-hand side, we have already said that 4 minus 4 is 0, leaving us just the x on the left-hand side. On the right-hand side, $10 - 4 = 6$, so the last line of the solution of the equation is:

$$x = 6$$

Read these notes through again to make sure that you understand what is going on.

Example 2

$$y - 5 = 17$$

Here, to get the y on its own, we need to add 5 to the left-hand side of the equation. Therefore, we must also add 5 to the right-hand side of the equation, in order to keep the equation in balance:

$$y - 5 + 5 = 17 + 5$$

$$y = 22$$

What about an equation with more than one lot of x on one side? Here, you have to use another technique. For example, look at this:

$$2x + 8 = 26$$

As before, isolate x by subtracting 8 from both sides, therefore the next line in the equation is:

$$2x + 8 - 8 = 26 - 8$$

$$2x = 18$$

However, we do not want $2x$, we want x on its own. In a situation like this, you should divide both sides by the **coefficient** of x (the number in front of the x). This gives:

$$\frac{2x}{2} = \frac{18}{2}$$

The twos on the left-hand side cancel, leaving the x. Cancelling means that the two in the bottom (**denominator)** of the fraction divides into the two in the top (**numerator)** of the fraction. This gives an answer of one, which is then multiplied by x, giving the x in the final line of the equation.

So the final line of the solution is:

$$x = 9 \quad (18 \div 2 = 9)$$

To summarise equations

1. Isolate the unknown (the letter such as x) one side of the equation. Aim to get everything else on the other side of the equation

2. When there is a coefficient of the unknown (number in front of the letter) divide both sides of the equation by that coefficient, and then show the final answer.

Remember, in algebra, when there is only one unknown, we just list it as the letter, for example, x or y and not $1x$ or $1y$. Mathematically, $1x$ is like saying 1 lot of 6. It is far better just to say 6. In the past we have had students who seem to have an emotional need to use the 1, arguing that whilst they are aware that the single unknown is a unitary quantity, they prefer to put the 1 in front. This looks immature on paper and is a practice to be avoided.

Solving equations with x on both sides

Although we used x in the title of this section, in an equation any letter can be used to represent the unknown (a, b, h, x, y, whatever you like). Solving an equation like this is not as daunting as it may at first appear. Take this example:

$$6y - 4 = 5y + 3$$

Again, our first job is to get all of the unknowns (letters) on one side of the equation, and all the numbers on the other. As a general rule, tackle the smaller quantities first. Here the -4 is a smaller quantity than the $+3$, so we add 4 to both sides. This undoes the effect of the -4:

$$6y - 4 + 4 = 5y + 3 + 4$$

We can now tidy this line up:

$$6y = 5y + 7$$

Now tackle the y by working on the smaller quantity. Let's subtract $5y$ from both sides, because this will eliminate the $5y$ on the right-hand side:

$$6y - 5y = 5y + 7 - 5y$$

Now tidy up this line, too. As you can see, the two lots of $5y$ on the right-hand side subtract to leave just the 7. The left-hand side subtracts to leave just the y. So the next line tells us the previously unknown value of y:

$$y = 7$$

You could also come across a similar type of equation that ends with a coefficient of the unknown. Here is an example:

$$5t + 7 = 3t + 17$$

Do exactly as before. Tackle the lower values first, so subtract 7 from both sides:

$$5t + 7 - 7 = 3t + 17 - 7$$
$$5t = 3t + 10$$

Now subtract $3t$ from both sides; this will isolate the t from the numbers, giving us:

$$5t - 3t = 3t + 10 - 3t$$

$$2t = 10$$

Here, you have a coefficient in front of the unknown. Remember to divide both sides of the coefficient by this number, in this case 2:

$$\frac{2t}{2} = \frac{10}{2}$$

$$t = 5$$

Equations with brackets

When dealing with brackets in algebra, you need to multiply out the bracket first. For example:

$$2(x + 5)$$

is expanded by multiplying the 2 and the x, to give $2x$. You then multiply the 2 and the 5, to make 10.

Let's look at this in the context of an equation:

$$5(m + 6) = 70$$

Expand the bracket by multiplying all of the terms in the bracket by 5:

$$5 \times m = 5m \text{ and } 5 \times 6 = 30$$

So the next line is:
$$5m + 30 = 70$$

Now treat it just like the equations we discussed earlier. Subtract 30 from both sides, to isolate the m:

$$5m + 30 - 30 = 70 - 30$$

$$5m = 40$$

Now divide both sides of the equation by the 5

$$\frac{5m}{5} = \frac{40}{5}$$

$$m = 8$$

You may be asked to solve an equation with two sets of brackets, like this:

$$7(x + 3) - 5(x + 1) = 4(x + 1) + 6$$

Here, it is important to expand all of the brackets first and then to collect together 'like terms':

$$7x + 21 - 5x - 5 = 4x + 4 + 6$$

Now, collect the 'like terms':

$$2x + 16 = 4x + 10$$

Now solve it in the same way as we did earlier:

$$2x + 16 - 2x = 4x + 10 - 2x$$

$$16 = 2x + 10$$

$$16 - 10 = 2x + 10 - 10$$

$$2x = 6$$

$$x = 3$$

Note this example down and then write in words the steps that we have made, as we have worked through the example. Then compare your notes with our notes.

This is the same example as above, but here we have put all of the notes in, to help your understanding. Make sure that you have a go at doing this first, before you read our notes.

$$7(x + 3) - 5(x + 1) = 4(x + 1) + 6$$

Here, it is important to expand all of the brackets first and then to collect together like terms:

$$7x + 21 - 5x - 5 = 4x + 4 + 6$$

Now collect like terms:

$$2x + 16 = 4x + 10$$

Now subtract $2x$ from both sides:

$$2x + 16 - 2x = 4x + 10 - 2x$$

Now collect like terms:

$$16 = 2x + 10$$

Now subtract 10 from both sides:

$$16 - 10 = 2x + 10 - 10$$

Again, collect like terms:

$$2x = 6$$

Now divide both sides by the coefficient of x (the 2):

$$\frac{2x}{2} = \frac{6}{2}$$

This then generates the final answer:

$$x = 3$$

As a check, substitute this into the original equation and see that it works:

$$7(3 + 3) - 5(3 + 1) = 4(3 + 1) + 6$$

$$7(6) - 5(4) = 4(4) + 6$$

$$42 - 20 = 16 + 6$$

$$22 = 22$$

So it is correct. Notice how the equals signs in an equation are kept directly below each other. This is a minor presentation matter, but it is important.

Searching the working

The following solution to an equation is wrong. Read through the solution and spot where the error has occurred:

$$4(x + 2) - 5(x - 1) = 3(x - 3) - 2$$

$$4x + 8 - 5x - 1 = 3x - 9 - 2$$

$$7 - x = 3x - 11$$

$$7 - x + x = 3x - 11 + x$$

$$7 = 4x - 11$$

$$7 + 11 = 4x - 11 + 11$$

$$4x = 18$$

$$\frac{4x}{4} = \frac{18}{4}$$

$$x = 4.5$$

Remember, this answer is *wrong*, but where is the error? Look carefully through the working and find the error, then write out the correct version of the solution to this equation. Then compare your notes with ours.

These were the first two lines of the calculation:

$$4(x + 2) - 5(x - 1) = 3 (x - 3) - 2$$

$$4x + 8 - 5x \underline{\mathbf{-1}} = 3x - 9 - 2$$

This is the line containing the error. We have emboldened and underlined the −1 to make it stand out. Here the expansion of the second bracket on the left-hand side is wrong. It should have been:

$$-5x + 5$$

so that term should have been +5 and not −1 as shown. So the correct solution should have been:

$$4 (x + 2) - 5 (x - 1) = 3 (x - 3) - 2$$

$$4x + 8 - 5x + 5 = 3x - 9 - 2$$

$$13 - x = 3x - 11$$

$$13 - x + 11 = 3x - 11 + 11$$

$$24 - x = 3x$$

$$24 - x + x = 3x + x$$

$$4x = 24$$

$$\frac{4x}{4} = \frac{24}{4}$$

$$x = 6$$

Solving equations with fractions

This usually causes panic amongst students. But it needn't: with a little thought, any problems can be overcome. Take this example,

$$\frac{x}{3} = 8$$

$\frac{x}{3}$ is actually a third of x, it could just as easily be written as $\frac{1}{3}x$. So ask yourself this, if a third of x is 8, what must be the whole x?

Clearly the answer must be 24, but we need a technique to work this out. The rule for equations with fractions is to multiply everything by whatever is in the denominator of the fraction:

$$\frac{x}{3} = 8$$

So here, multiply everything by 3:

$$3 \times \frac{x}{3} = 3 \times 8$$

which leads us to:

$$\frac{3x}{3} = 24$$

Remember, the 3s on the left hand side cancel out. Think about it:

$$3 \div 3 = 1$$

and then

$$1 \times x = x$$

which leaves us with:

$$x = 24$$

Another example

$$\frac{3x}{4} = 12$$

As before, multiply both sides of the equation by 4, because 4 is in the denominator:

$$4 \times \frac{3x}{4} = 4 \times 12$$

Notice that on the left-hand side the fours cancel, leaving $3x$,

and since

$$4 \times 12 = 48$$

the next line of the working is:

$$3x = 48$$

We have already seen how to solve equations of this type. Notice how we have manipulated the first equation and changed it from an unfriendly form, into a form that we already know how to manipulate. Solve this equation now, before you move on and then compare your solution with the solution that we have worked out below.

If you have not tried solving the equation above, please do so now, before you read our solution.

Since $3x = 48$ we need to divide both sides by 3, because this is the coefficient of x:

$$\frac{3x}{3} = \frac{48}{3}$$

As before, the threes on the left-hand side cancel, leaving the x. On the right-hand side:

$$48 \div 3 = 16$$

so $x = 16$

Problem solving and equations
The main use of these techniques is in problem solving. For example, if the length of a rectangle is 4cm more than its width, and its perimeter is 24cm, find its length.

In a question like this, always draw the diagram first.

$x + 4$

x

Fig. 68

Since we don't yet know the width of the rectangle, let's call it x. Since the length of the rectangle is 4cm more than its width, we will call this $x + 4$.

We know the perimeter of the rectangle (the distance around its boundary) is 24 cm and this perimeter is made up of 2 widths and 2 lengths. Mathematically we can write this as:

$$2(x) + 2(x + 4) = 24$$

Here we have an equation similar to those that we worked on earlier:

$$2x + 2x + 8 = 24$$

$$4x + 8 = 24$$

$$4x + 8 - 8 = 24 - 8$$

$$4x = 16$$

$$\frac{4x}{4} = \frac{16}{4}$$

$$x = 4\text{cm}$$

Since x is the width, the width is 4cm, but we were asked for the length. Since the length is 4cm more than the width, the length must be 8cm.

This example shows that you sometimes need to set up your own equation and then solve it. Use the practice questions at the end of the chapter to help you consolidate these skills.

Squares and square roots

It is important to know how to 'undo' a square number. For example, 4 is a square number:

$$\sqrt{x} = 4$$

This type of equation reads as 'the square root of x equals 4'. To undo an equation like this, we need to square both sides of the equation. That is, we must multiply both sides of the equation by themselves. Why? Because $\sqrt{x} \times \sqrt{x} = x$, the quantity that you want.

Since we always treat equations like a balance, we must also square the other side of the equation. So here,

$$x = 16$$

Changing the subject

In mathematics it is also known as

the **transposition** of formulae

or the **transformation** of formulae

Remember, the Latin word 'formula' is singular, and the plural is 'formulae'.

Changing the subject means rearranging a formula to make any one of the letters appear on its own, and moving everything else to the other side of the formula. To achieve this, we need to use all the rules reviewed in this chapter for solving equations.

Making x the subject

Make x the subject of the following formula:

$$f = t + 2x$$

This is an equation, so treat it in the same way as the equations that we dealt with earlier. We want x to be the subject, so isolate x by subtracting t from *both sides* of the equation:

$$f - t = t + 2x - t$$

Here, the t's on the right-hand side subtract from each other, leaving:

$$f - t = 2x$$

Now divide everything in the equation by 2 (the coefficient of x):

$$\frac{f - t}{2} = \frac{2x}{2}$$

Now the twos on the right-hand side cancel, leaving as the final answer:

$$x = \frac{f - t}{2}$$

Let's look at another example, using brackets. Make x the subject of this formula:

$$v(x + v) = y^2$$

First of all, expand the brackets:

$$vx + v^2 = y^2$$

Now subtract v^2 from both sides. This will eliminate the v^2 on the left-hand side of the equation and leave us with vx on the left-hand side:

$$vx + v^2 - v^2 = y^2 - v^2$$

This then gives $vx = y^2 - v^2$

Now look at the left-hand side. It is a **product** of two numbers: vx is a number produced by multiplying the numbers v and x.

To undo a product we need to *divide*, and to divide by the coefficient of the product. This is exactly the same process that we dealt with in equations. Remember when we had equations such as $2x = 4$ and when we divided both sides of the equation by 2? This was because the coefficient was x. In fact $2x$ is a product, of $2.x$ (in maths $2.x$ means $2 \times x$).

$$\frac{vx}{v} = \frac{y^2 - v^2}{v}$$

Here the two vs on the left-hand side cancel, leaving us with:

$$x = \frac{y^2 - v^2}{v}$$

Whilst this is an acceptable answer, you need to be aware of one extra mathematical technique. In the equation, on the right-hand side, v is a common denominator, so the answer could be written as:

$$x = \frac{y^2}{v} - \frac{v^2}{v}$$

In circumstances like these, we can tidy up:

$$\frac{v^2}{v}$$

In mathematical terms, this is a **quotient**. In other words the v^2 in the numerator is being divided by the v in the denominator. Remember, when dividing, we subtract the powers, so:

$$v^2 \div v = v^{2-1} = v$$

Therefore the final line of the equation:

$$x = \frac{y^2}{v} - \frac{v^2}{v}$$

becomes

$$x = \frac{y^2}{v} - v$$

Changing the subject of formulae involving fractions

In some formulae, you will be presented with an equation where the x is wrapped up in a formula. Again, as we showed when dealing with equations, whenever you have a fraction, to undo or get rid of the fraction, multiply both sides of the equation by whatever is in the denominator of the fraction. For example:

$$\frac{x}{p} = a$$

Here, we have p in the denominator, so multiply both sides of the equation by p; the resulting effect will be for the p's on the left-hand side of the equation to cancel out:

$$p \times \frac{x}{p} = p \times a$$

This gives the final answer:

$$x = pa$$

Occasionally you will see formulae with fractions also wrapped up with brackets. For example:

$$\frac{x}{m} = (w - y)$$

Again, the first thing we must do is to eliminate the fraction, so multiply both sides of the equation by m, because m is in the denominator:

$$m \times \frac{x}{m} = m \times (w - y)$$

As usual, the m's on the left-hand side cancel, leaving the x, so the final answer is:

$$x = m(w - y)$$

What happens when x is in the denominator of the fraction?
As with any fraction in an equation, to eliminate the fraction, simply multiply everything in the equation by what is in the denominator, in this case the x. Look at this example:

$$\frac{m^2}{n^3} = \frac{v^2}{x}$$

Multiply both sides of the equation by x:

$$x \times \frac{m^2}{n^3} = x \times \frac{v^2}{x}$$

Here the x's on the right-hand side of the equation cancel out. This then gives:

$$\frac{xm^2}{n^3} = v^2$$

but we still have a fraction and we do not have the x on its own. As before, multiply both sides of the equation by n^3:

$$n^3 \times \frac{xm^2}{n^3} = n^3 \times v^2$$

As before, the n^3 terms on the left-hand side cancel, so the equation now becomes:

$$xm^2 = n^3v^2$$

Now we need to recognise that the left-hand side is a product, so divide both sides of the equation by m^2

$$\frac{xm^2}{m^2} = \frac{n^3v^2}{m^2}$$

Now the m^2s on the left-hand side cancel, leaving us with the final answer of:

$$x = \frac{n^3v^2}{m^2}$$

Read through this working again and make sure that it is clear to you.

Formulae with x^2 and negative x terms
What happens when there is an x^2 value? Let's use this example:

$$bx^2 = f$$

We have already recognised, from the work above, that the left-hand side of the formula is a product, therefore we undo the product by dividing both sides of the equation by whatever is the coefficient of x, in this case, b:

$$\frac{bx^2}{b} = \frac{f}{b}$$

As before, the bs on the left hand-side of the equation cancel out, leaving us with:

$$x^2 = \frac{f}{b}$$

From here, we need to eliminate the x^2 by reducing it to x.

Since x is the square root of x^2, we simply take the square root of both sides of the equation:

$$\sqrt{x^2} = \sqrt{\frac{f}{b}}$$

giving us:

$$x = \sqrt{\frac{f}{b}}$$

What happens when we have a negative x?
Here we need to rearrange the formula to make x a positive term, for example:

$$m - nx = t$$

If we add nx to both sides, the effect is to make the nx term positive.

$$m - nx + nx = t + nx$$

Here the nx terms on the left-hand side cancel out, so:

$$m = t + nx$$

Now subtract t from both sides:

$$m - t = t + nx - t$$

Now the ts on the right-hand side cancel out:

$$m - t = nx$$

Here nx is a product, so divide both sides by n:

$$\frac{m - t}{n} = \frac{nx}{n}$$

Here the ns on the right-hand side cancel, leaving us with

$$x = \frac{m - t}{n}$$

Again, read through these notes and make sure this is clear for you.

Tutorial

Practice questions

A. Solve the following equations:

1. $x - 5 = 7$
2. $y + 7 = 12$
3. $2x + 7 = 3x + 2$
4. $3t + 1 = 22$
5. $5x - 4 = 4x + 5$
6. $4 + 8x = 9x + 1$
7. $6x - 3 = 5x + 4$
8. $4(x - 2) = 3(x + 1) - 6$
9. $5(x + 3) = 9(x + 1) - 14$

10. $\dfrac{4}{x} = 7$

11. $\dfrac{9}{x} = 12$

12. $6 + \dfrac{3}{x} = 9$

13. $\dfrac{x}{7} = 6$

14. $\dfrac{2x}{7} = 10$

B. Rearrange the following formulae, to make x the subject:

1. $x + m = c + b$
2. $xk - j = v$
3. $t(m + x) = v$
4. $n(b - x) = h + f$
5. $\dfrac{x}{(a+b)} = (m + n)$
6. $d(a + x) = y$
7. $\dfrac{y}{x^2} = t$
8. $\dfrac{x^2}{y} = 2x$
9. $fx^2 - d = f^2$
10. $\dfrac{m(x+d)}{a} = z$

C. Find the **error** in this working:

$$2x + 5 = 3x + 2$$
$$2x + 5 + 2x = 3x + 2 + 2x$$
$$4x + 5 = 5x + 2$$
$$4x + 5 - 4x = 5x + 2 - 4x$$
$$5 = 5x + 2$$
$$5 - 2 = 5x + 2 - 2$$
$$5x = 3$$
$$x = \frac{3}{5}$$

Remember, this answer is *wrong*, but where is the error?

D. Find the error in this working. Rearrange the formula to make x the subject:

$$\frac{4}{x} + 4 = c - d(1+f)$$

$$\frac{4}{x} + 4 = c - d(1+f) - 4$$

$$\frac{4}{x} = c - d(1+f) - 4$$

$$\frac{4}{x} \times x = xc - d(1+f)$$

$$4 = xc - d + df$$

$$\frac{4}{xc - d + df} = x$$

Again this is *wrong*, but where are the mistakes? Discuss this with another person and make sure you both agree.

Seminar discussion

1. Using symbols to solve problems is essential to becoming what Robert Reich calls 'symbolic analysts'. Read his book *The Work of Nations*. Do you consider his view on symbolic analysis correct?

2   Mathematical education for too many people means number work. We have to move away from the idea that it is acceptable to be bad at maths; we have to treat it as seriously as we treat literacy. Many people proudly state that they could never do maths at school, yet few admit to being illiterate. Why is this?

Practical assignment

Set up a spreadsheet on a computer. Use the facility of 'formulae' in the spreadsheet to make calculations relevant to your course.

Study tips

1. Discuss your use of algebra with another person. Explain to them how to solve equations and how to rearrange formulae; this will aid your understanding.

2. Make sure that you understand negative numbers and fractions, when working with algebra; if in doubt, ask your tutor for advice.

Spreadsheets

One-minute overview — In this chapter we will look briefly at spreadsheets. We will be examining:

▶ how to open a workbook

▶ the names of the different parts of the workbook

▶ how to use a spreadsheet for basic mathematics

In this chapter we will use Microsoft Excel as our exemplar spreadsheet. This is not the only spreadsheet software that is commercially available but, owing to the dominance of Microsoft Office as the main software used by most people, it seems sensible to use this as our exemplar. We think you will be impressed with the power of the spreadsheet and how it can help you (screenshots courtesy of the Microsoft Corporation and Open Office). If you do not own the Microsoft Office package then a suitable and free alternative called Open Office can be downloaded from the internet. Just go to a search engine like Google, type in Open Office and then press Google search.

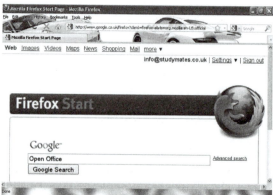

Fig. 69

Then you will get a page looking similar to this:

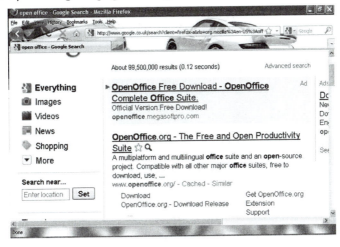

Fig. 70

Now click on the download here and your screen will now look like this:

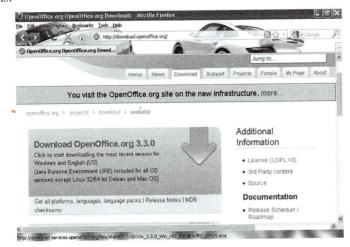

Fig. 71

Follow the instructions on screen and you will download the package for free. There is therefore no need for anyone unable to buy a spreadsheet package to be left without a spreadsheet. The free Open Office Spreadsheet is called scalc. You can also save the Open

Office Spreadsheet as a Microsoft Office Excel spreadsheet.

When you open an open office spreadsheet it will look like this:

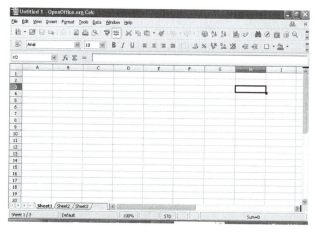

Fig. 72

When you press File and save as, your screen will look like this:

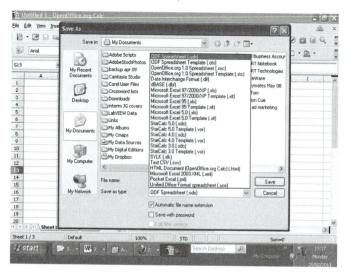

Fig. 73

Here you can see the Microsoft Excel options and so can save your free Open Source Spreadsheet in a format that others can use on their machines when they are using Excel.

However, for the rest of this chapter we will stay with Excel.

When you open Excel you will see a workbook which is made up of worksheets.

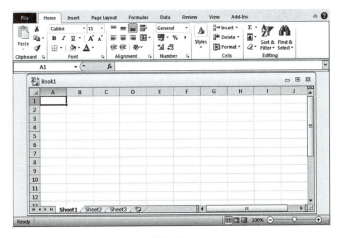

Fig. 74

The workbook is made up of small rectangles, which are called cells. They are known by the letter across the top of the worksheet and the number down the left hand side.

How to magnify or reduce your display

If you need to magnify or reduce a worksheet then click on the view tab on the ribbon.

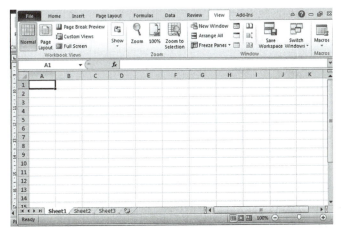

Fig. 75

We have shown it in the figure. Now click on the zoom magnifying glass.

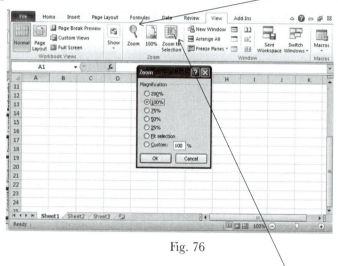

Fig. 76

This will then bring up different sizes to either increase or decrease the size of the spreadsheet. Look at the zoom to selection button on the tab.

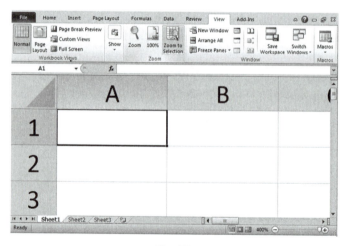

Fig. 77

You can now choose the most appropriate size for your needs. Changing the magnification does not affect printing. Sheets are printed at 100% when you use this facility.

How to move and scroll through a worksheet

In order to move between cells on a worksheet, click any cell or use the arrow keys on your keypad. When you move to a cell, it becomes the active cell. Here the active cell is E8.

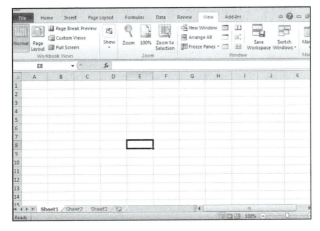

Fig. 78

To see a different area of the sheet, use the scroll bars.

Inserting the same information in several worksheets at once

Let's say you wanted to fill in data for three shops for three weeks.

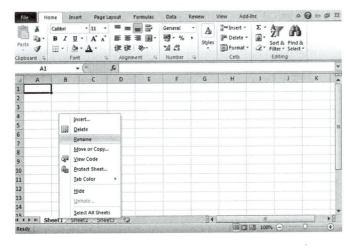

Fig. 79

Here we have right clicked on the tab saying worksheet 1 and clicked on rename. We will now rename it and rename the other two tabs, so the first tab will be for our

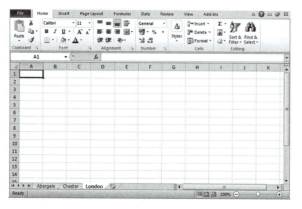

Fig. 80

Abergele shop, the next for our Chester shop and the third for our London shop. So now the worksheet looks like this.

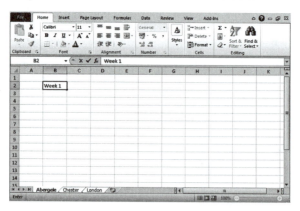

Fig. 81

You can see that the tabs now say Abergele, Chester and London. We want to enter data that will apply to all of the worksheets. We would like to enter *week 1, week 2* and *week 3* at the top of all three worksheets.

Click on the tab for the Abergele worksheet. Hold down the CTRL key whilst you click on the Chester and London tabs.

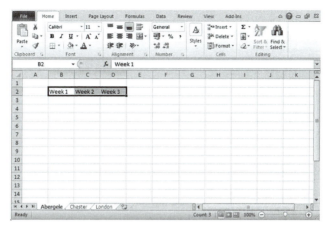

Fig. 82

If you click on the other tabs for Chester and London you will find that you have also inserted Week 1, 2 and 3 onto these worksheets as well, thereby saving both time and effort.

Now we will enter titles from our Studymates list onto the worksheet. If you look at the worksheet as it stands, there is clearly not enough space in the A column to write the titles of books.

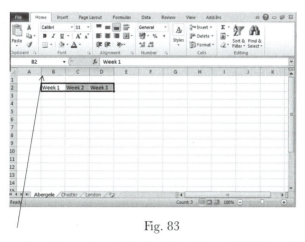

Fig. 83

Click on the tabs again, whilst holding down CTRL and then click on the line between the A and B columns. You can then drag this line to the right and it will widen the A column. Notice that because you have clicked on the tabs (whilst holding down CTRL), it also widens the A column for the Chester and London tabs.

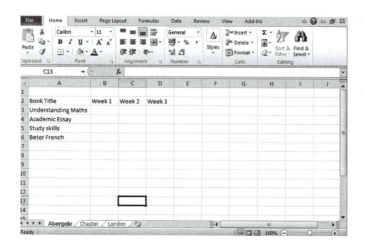

Fig. 84

In Figure 84 we have started entering the Studymates titles down the left-hand side (Studymates books and Aber books are published by the same company). You will notice that there is a spelling error in A6. All we need to do is double click on that cell and we can edit it. Remember that because we have the tabs pressed down at the bottom of the worksheet, that spelling correction automatically happens for the Chester and London shop as well.

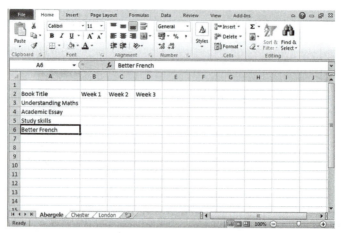

Fig. 85

Moving cell contents

If you wish to move the cell contents of one or more cells to a new location, then select the range to be moved by dragging across the cell and then drag on one of the boundaries. (Make sure that you keep away from the corner handle.)

Freezing panes

This workbook will be used to monitor sales of Studymates titles and Aber/Mr Educator titles when they are added at a later date, (see *www.aber-publishing.co.uk*)

Therefore it is important to be able to add more data. To do this we need to be able to freeze the pane and scroll the data.

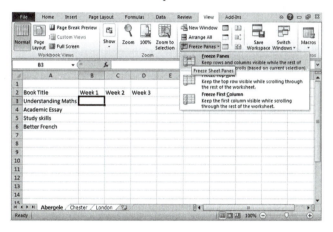

Fig. 86

To freeze the panes we have selected B3, then clicked on the view tab and then freeze panes. This means that the book title list will stay where it is and the week 1, 2, 3 etc. columns will move to the left and go out of view. Notice we have left the tabs for all three shops depressed so this will freeze the panes on all three worksheets.

Occasionally, you will may that your needs change and that you may need to unfreeze the same pane and then either leave it or freeze another pane. Repeat the process above but when the Freeze panes tab is pressed then an option to unfreeze appears. Your screen should look like Figure 87.

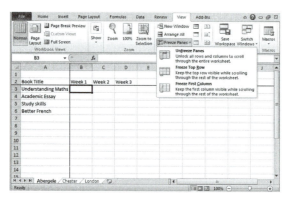

Fig. 87

Click on the unfreeze pane and you then have a worksheet to work on as you wish.

Entering simple formulae

Spreadsheets are designed to calculate. It is possible to enter numbers in some cells and then enter formulae in other cells that tell the spreadsheet what operation to make on the data you have entered.

Usually a formula is made up of a combination of cell references and numbers and specially written spreadsheet functions like *sum* or *average*.

Whenever you want to write a formula you must start with an equal sign (=).

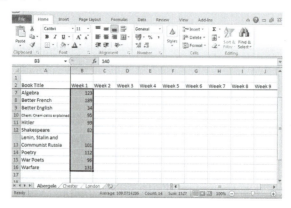

Fig. 88

Notice that we have highlighted the data. Click in cell B17 and then click the sigma button (Σ).

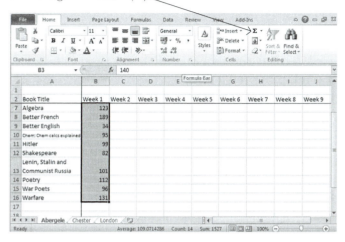

Fig. 89

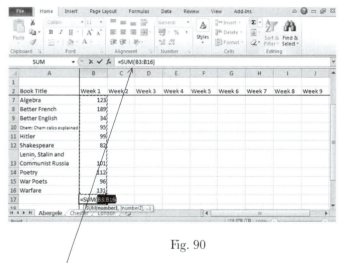

Fig. 90

Your screen should now look like this (Figure 90). Look at the formula in the formula bar. It is telling the computer to add up (sum) all of the numbers from B3 to B16 and to put the answer in B17.

You should have found that for week 1 the Abergele shop sold 1527 books. We had the tabs down for Chester and London so this data will also appear on those pages, making it easier to compare sales between the shops. This opportunity to compare sales data is vital in business. Abergele is a town on the North Wales coast so we would not expect to sell as many books as the Chester shop. Chester, whilst being a beautiful city and a great place to live is not as big a city as London and therefore we would not expect to sell as many books in Chester as in London.

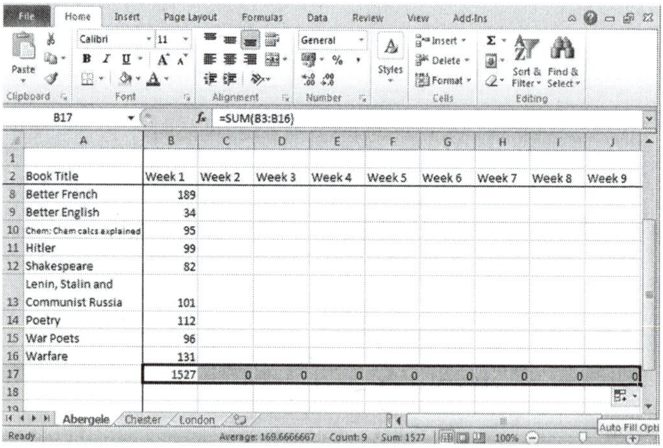

Fig. 91

In this scenario we will be putting in the sales figures for each of the titles in each subsequent week under the relevant column header. So the sales for week 2 go under week 2, those for week 3 will go under week 3 and so on. Each of these columns needs to have a formula to add them up. An easy way is simply to drag the formula across from the first column as demonstrated here. You will see the spreadsheet shows zeros all the way across but if you put two numbers for sales under week 2, if you have dragged the formula properly (drag on the bottom right hand corner), then the sum of these two numbers (the numbers added together when they are in the same column) should appear in line 17.

Here, in the next screenshot of a spreadsheet, we have completed a small amount of computer gimmickry to show what is actually happening 'behind the scenes'. The computer has a formula in B17 that tells it to add the contents of all of the cells from B3 to B16 inclusive (i.e. including B3 and including B16). By dragging the formula across the computer has been programmed to recognise that you want this formula in the other cells. But the computer is also programmed to recognise that in Column C, the formula will not work because the formula is 'add everything from B3 to B16'. So when the program registers the mouse over column C, the formula is updated and C replaces B. As you go across the columns, this process continues until eventually you have a spreadsheet with these formula:

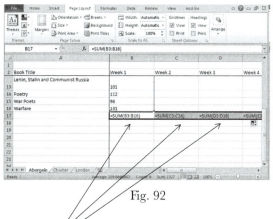

Fig. 92

The spreadsheet automatically updates the formula as it moves from left to right across the worksheet. So, if you put in numbers they will automatically add for you.

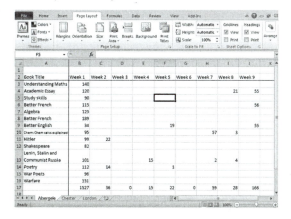

Fig. 93

You can see here that we have started doing exactly that and this is why spreadsheets are so powerful and worth understanding.

The average

Microsoft use the term *average* to stand for the mean of a set of data. Here in cell B18, we have put the formula $=$ average(

Then we have dragged the cursor from B3 to B16 and put the closing bracket in and pressed enter. This function then calculates the mean for the whole column.

Formatting

You will need to format certain aspects of your spreadsheet. Since the software originates in the USA, it is often the case that the default settings are for that country.

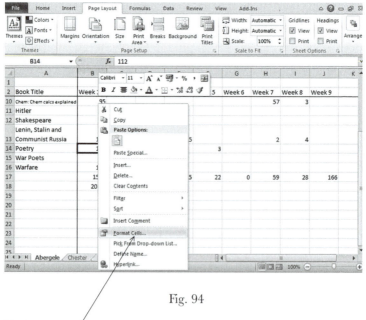

Fig. 94

Click on format and you should find that this happens:

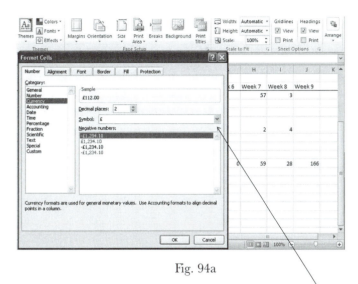

Fig. 94a

If you are not US based you will need to pick your local currency, here since we are in the UK we have picked pounds. Be aware that more and more UK businesses are also pricing in Euros so click on the down arrow here and make sure you know where the € option is located.

If you click on the alignment tab you also have the option of wrapping text in cells. This makes the spreadsheet tidier. It is such a powerful tool that many young children below the age of 10 are being taught how to use a spreadsheet. You should now have enough information to be able to start using a spreadsheet.

Tutorial

Set up a spreadsheet to monitor your expenditure on food. Make a separate column to show how much is spent on takeaways and how much is spent on fresh food. Set up a separate column that is 10% of the value you spend on takeaways and resolve to spend it on fresh food from now on, 10% is 0.1 so calculate 0.1 x the amount spent on takeaways. You may be surprised how much extra food you can buy for your money.

Seminar discussion

Explain to another person why it is important to track your personal spending. Make sure you mention that spreadsheets allow you to spot financial leaks and to plug them. For more on personal finance see *Back to the Black: How to get out of Debt and Stay out of Debt* from Aber Publishing 9781842851418.

We have often found that when taking students through this exercise that there is an element of surprise. One young man proclaimed 'I can't afford to keep going out to clubs four nights a week.' We asked him how he had made that judgement and he said that 40% of his part-time income had been spent paying entry fees to clubs. We encourage you not to make judgements on people regarding what they do with their money, simply to empower them to understand the implications of their actions and to encourage them to think ahead.

Study tips

Make sure you are clear about using the spreadsheet package before you offer to do anything in your workplace. Nothing can destroy your confidence more than setting yourself up as an expert but being unable to deliver.

Appendix

One-minute overview – In this chapter we have diverted from the usual format in a mathematics book because we want to bring the whole thing together and to challenge your thinking about mathematics and about your future.

In this chapter we will look at:

▶ Personal development profiles.
▶ Reviewing learning and the review rules.
▶ Fitness.
▶ Reading.
▶ Why mathematics counts.
▶ Problem solvers.

Personal development profiles (PDPs)

One of the new developments that is taking place in many industries is the use of PDPs. This is a document that can take many formats but is basically a record of all training you have completed in your job and a plan of the training you intend to do in the future. We think that the best way to think of the PDP is as a record of:

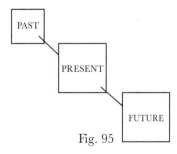

Fig. 95

An employer will employ you because of the knowledge and skills you can bring to the organisation. You are being employed to sell your knowledge and skills to the company or organisation that employs you.

Philosophically this is too much of a cultural challenge to some people and we do know of large training organisations where the development of the PDP is being developed amongst trainees as a voluntary option. We believe this is an error of judgement. We know of cases where interviewees have attended for an interview only to be asked at the entrance for their PDP. Those that did not have one were given an expenses form and not allowed to enter the building.

In a knowledge-based economy, you are selling your knowledge to your employer. If you have no new knowledge, why should they continue to buy your labour?

Bad managers

We know of a case in a company in the West of England where a manager came into work one day and was asked to see the director. She went to take her coat off and was asked not to do so and to see the director immediately. This she did and still wearing her coat was told that after 15 years of service to the company she was no longer required. This was a personally devastating experience for the manager concerned. She had just returned from a successful sales trip in Europe and was extremely pleased to have secured major orders for the company. All of this counted for nothing, she was escorted from the building and her personal belongings were packaged up and sent to her via Royal Mail. This was harsh and in our view poor management. The company were then left with the task of finding and training another person. Although UK law clearly states that a person can only be made redundant under certain circumstances, one being that the job has disappeared in company restructuring, the company got round this by changing the job title and did appoint a new person, a man. We believe that this was a case of discriminating against an older woman in the workplace.

This person's real problem was that she had not been on a course for the full length of time that she had worked for the company. So in 15 years she had not learned anything new. Therefore her line manager's perception was that she had nothing further to offer. He was quite wrong, she had excellent relationships with her foreign customers and had kept sales up in a difficult market but this counted for nothing.

Freelancers experience this type of discrimination on a regular basis. There is one well-established company in the UK where we know that a freelancer was phoned and told by his manager that he needed to make certain changes to his work. This he did and honestly thought all was well. The female manager then phoned him up three weeks later from her home in Manchester and without warning stopped his work.

We know of a company in the South-West of England where a director had a full stand-up row with an employee. The employee was booked to be in two meetings at the same time, despite having told the director previously. The employee then committed the 'crime' of asking to see the minutes of the meeting, after the meeting was over. He was astonished to hear the director say they were not recording any minutes of any meetings ever. The project they were on was reported to have lost nearly £4 million, hardly surprising when you have that approach to management.

We present these cases because we believe that there are a lot of inadequate managers in industry. These are people who have power over workers and they themselves feel vulnerable and will remove people they see as threats. Therefore we believe that it is in your interest to have a well-developed PDP. In the event of you working for an inadequate person like this, you are better off having a good PDP because it empowers you to move to find a better employer.

How to complete a PDP

Step one
Make a personal written statement of where you are now in your career.

Example
I am now employed full time as an apprentice in the engineering maintenance department of XYZ Company. I am married/divorced/engaged/single/ in a relationship and have 0/1/2/3 etc children. My hobbies are X/Y/Z and I enjoy A/B/C.

This should give you a written statement that represents your life as it is now.

> **Remember it is not where you come from that is important; it is where you are going that is important.**

Step two

Write down a vision of where you want to be in three years' time or in five years' time.

Example

In 20__ I will be the manager of XYZ, I will run the department efficiently and we will make/do/sell.......................
My personal life will be A/B/C and I will be/have/get.............

In your vision make it as real as possible. What type of house will you live in? What colour paint will go on the walls? What will the furnishing look like? What will the garden (if you have one) look like? What type of car will you drive? Smell the leather seats in your car; admire the house that you will be living in.

Step three

This is the goal setting to help you get from where you are to where you wish to be. You need to make a SWOT analysis. SWOT stands for:

Strengths
Weaknesses
Opportunities
Threats

List the **strengths** that you have. We all have them – for example:

► Are you a good communicator?

► Do you ensure that people clearly understand what you are saying or writing?

► Are you a good manager? (Many women who have taken time off work to raise children fail to recognise that running an efficient home where the family is happy is the sign of a very good manager.)

Weaknesses: be honest with yourself, what are you not so good at? This book will have helped you with your mathematics but what about taking it to a higher level? What about joining a night school or the Open University? (If you are not resident in the UK, you can still join the UK's Open University or the country where you live may have their own version of the OU. We highly rate the Open University, the author has two OU degrees and his partner has one OU degree).

This is where you need to plan the courses you intend to take. You should consider at least one course per year; this is what employers are looking for. If you cannot get time off work and if your employer will not support you, then should you really be working for that employer? There is an old management saying 'What if you train them and they leave?' To which the response is, 'What if you don't and they stay?

Opportunities: what opportunities exist in your location for training? Think carefully before you dismiss these opportunities. Many jobs are never advertised; it is often the case that training organisations point trainees towards new opportunities. One thing to consider is starting your own business. It is *not* easy. We know many people who have started businesses and been successful but only after many years of solid hard work. So please do not think this is an easy option. We certainly want you to consider further work in mathematics and we will discuss this later in this chapter.

Threats: this is the scary one. What threats are there to your livelihood? We know of cases of people in their forties who have lost their jobs and are totally unprepared. We did hear of one man who had been earning £80 000 per annum in sales in the 1990s but was now on income support. In five years he had applied for less than 50 jobs. That's less than ten a year. He should have been applying for 50 per week.

Be realistic, there are threats to your income, what are they and how can they be minimised? Only you can answer this *but* we suggest that one way of minimising them is to increase your education and make sure your knowledge is up to date.

Step four
Now is the time to start taking action. You are responsible for the

future you create for yourself. You are where you are in life because of the decisions and actions you have taken in the past. People who are successes are people who take that responsibility on themselves. They do not blame outside circumstances, they do not blame a lack of money or technology or some other aspect.

> There is an interesting case of the founders of a well-known international business. They sold the company and went their separate ways as rich men. Both invested in subsequent businesses and both lost a lot of money. One partner took responsibility for himself and recognised he had made mistakes. He created an action plan that would change his life. He became rich (again) and made even more money than before. The other partner blamed the seasons; he blamed the weather in fact he blamed everything else except himself and adopted the role of victim. He consistently failed to see that he alone is responsible for where he is in life and that only he can improve his situation. Until he makes that change in thinking, he will never change his state in life.

You are not a victim; you are a superior person who cares about his or her future.

You now need to create an **action plan**. This is a set of targets that will move you closer to your vision. Ask yourself: what can I do today that takes me closer to the vision? Most commentators say that you should aim to do five things every day that take you closer to your vision for your future.

Make a 90-day action plan

▶ Where will you be in 90 days?
▶ What changes will you have made to your life?
▶ How much closer will you be to having the life you truly desire?

Break that down into 30 days. Where will you be in your plan in about 30 days?

Break the 30 days into a weekly plan and from that a daily plan. Remember the sign of a superior person is that s/he can take action. It means that they will focus on their goals and achieve them.

Think of athletes like Dame Kelly Holmes, think of her action planning. She will have gone through a similar thinking process to achieve her wonderful successes. Don't forget she was plagued with injuries, it would have been very easy for her to give up and say it was not possible to achieve what she had set out to achieve because of external factors beyond her control. But she didn't do that. She focused on her goals, she persevered and was hugely successful in her field. This is a woman who started her career in the army. She didn't come out of university armed with degrees and certificates. She set her goals and worked quietly and methodically towards achieving them – you can do the same.

In your action plan for your PDP you need to include readings and a fitness programme. Many people sneer at reading but it is the best way to get accelerated experience. For a small price you get access to the world's experts in different subject areas. If you read two books per week, in a year you will have read 104 books. This will most certainly put you ahead in your chosen field. In surveys in the USA, this is what the top 100 executives do every week. They spend two hours per day reading and studying because they know this is what will drive them to the top in their quest to develop their minds. You also need to plan regular exercise. If you are in a job where you get regular exercise then this is not a problem but many of us spend hours behind a desk and that is not healthy. How can you expect to be at the top of your field if you are unfit and sluggish?

Reviewing learning
What has all of this got to do with mathematics? In the same way that you review your goals you have to review your understanding of the mathematics and build it into a programme of personal development.

Review rules
▶ 10 minutes
▶ 1 hour
▶ 1 day

- ▶ 3 days
- ▶ 1 week
- ▶ 3 weeks
- ▶ 1 month
- ▶ 3 months
- ▶ 1 year
- ▶ 3 years

This is the review programme you should adopt for any new learning. The point to understand is that whenever you learn something new, it goes straight into your long-term memory and then it goes straight out again. By reviewing the work you have learned, your brain changes and you are reinforcing the neural pathways to ensure that you have learned the material.

Professor Robert Winston once used a brilliant metaphor for learning. He likened learning to building a rope bridge across a canyon. Think about how you would do this. First of all, you would need to get a rope across and securely fixed. Using that rope you could possibly cross the canyon but it would be a little dangerous to say the least. Then perhaps a second rope could be crossed over the canyon. You could then balance across both of these ropes. This means the crossing would be easier than with just one rope but it would still be pretty precarious. The next stage would be to lay down some wooden slats and build a walk-able bridge. This is how to think of learning. You are building a bridge in your mind between new material that is entering your mind and other already learned material. There is evidence that as we learn the brain does change and pathways do open up. The great news is that as you use this particular neural pathway more and more, the pathway becomes established and the information gets lodged in your long-term memory.

We recommend that you look at Tony Buzan's *The Mind Map Book* (0-563-37101-3 BBC Books) for more help on learning. Dr Buzan is a UK expert on learning and developed the mind map approach.

Why mathematics counts

Mathematics is important in your career. Having mathematical ability means that in terms of your ability to think you are ahead of most people. It means you have the thinking ability that leads you to be able to apply your knowledge to solving problems. This is the area where most mathematics is needed in industry. As Sir Arthur C Clarke says in the foreword to this book:

> *Those who cannot relate to percentages, fractions, probabilities, and statistical projections would find themselves quite unable to navigate the knowledge society or to survive in the knowledge economy.*

There is a new type of job being developed in industry, which is broadly called a strategic thinker. This is what may have been called a trouble-shooter in the past and these jobs are increasing. They are not just for people who have been to university. They exist at all levels within an organisation, so there is real benefit for you in terms of your career, but there is also another benefit.

Mathematics is an interesting subject in its own right. It actually serves two roles, one as the maidservant of other subject areas but although it may seem strange to hear an author say this, mathematics is of great interest in its own right to millions of people. Think about all the puzzles you have ever had fun with; they are largely mathematical. Lewis Carroll, the author of *Alice in Wonderland*, was in real life Charles Dodgson the mathematician. He also wrote some very famous puzzles called '*Tangled Tales*' as an entertainment. So whilst we have concentrated on the functional aspects of mathematics here we want you to be aware that there are aspects of mathematics and mathematical thinking that are intellectually challenging and a delight.

We would like to leave you with this puzzle:

> *At the start of the First World War, soldiers wore leather caps but then steel helmets were introduced. The effect of this was that head injuries went up not down, why?*

Answers to Questions

Chapter 1

Practice questions *(page 14)*

1. $^-3$ 2. $^-4$ 3. $^-7$ 4. $^-68$

5. $^-100$ 6. $^-100$ 7. $^-10$ 8. $^-6$

9. $^-480$ 10. It is wrong, the operations should be:

$$^-6 \times {}^-9 = 54$$
$$54 + 7 = 61$$
$$61 + 5 = 66$$
$$66 \div 6 = 11$$

Progress questions *(page 16)*

1. (a) 2.3536   (b) 1560

2. (a) $^-20$ (b) 15

3. (a) $^-4$ (b) $^-18$

Chapter 2

Practice questions *(page 29)*

1. $\frac{5}{9} = \frac{10}{18} = \frac{15}{27} = \frac{20}{36} = \frac{25}{45} = \frac{30}{54}$

2. $7\frac{5}{8}$

3. $2\frac{11}{12}$

4. $9\frac{7}{30}$

5. $16\frac{31}{60}$

Practice questions *(page 31)*

1. 375   2. 160   3. £100   4. £60
5. $\frac{1}{5}$ did not vote

Progress questions *(page 32)*

1. $\frac{2}{3}$ $\frac{4}{6}$ $\frac{6}{9}$ $\frac{8}{12}$ etc
2. (a) $\frac{11}{24}$ (b) $13\frac{5}{8}$ (c) $9\frac{1}{4}$
3. (a) $\frac{1}{4}$ (b) $\frac{1}{6}$ (c) $3\frac{1}{4}$
4. (a) $\frac{3}{8}$ (b) $\frac{10}{27}$ (c) $14\frac{7}{12}$
5. (a) 6   (b) 4   (d) $4\frac{1}{5}$
6. $\frac{1}{6}$

Chapter 3

Practice questions *(page 36/37)*

1. 24   2. 1300
3. 4% of 500 is 20   5% of 400 is 20   so they are both the same
4. £18 720   5. £81 600   6. £292.50   7. 79 750
8. £18 656   9. £16.50   10. £15 300

Practice questions *(page 48)*

1. 8%   2. £27 000   3. £13 650   4. £128
5. £1.05   6. £1.12   7. $c = 3n$, where n is the
                                        number of stamps and
                                        c is the cost
8. £1040   9. £160 000 for the 40-year-old and
                       £240 000 for the sixty-year-old
10. A gets £20 000 and B gets £30 000

Progress questions *(page 49)*

1. 126   2. 2310   3. 704   4. £2.40

5. £1.80

Chapter 4

Progress questions *(page 61)*

1. (a) 5   (b) 700   (c) 20 000

2. $\frac{9}{20}$

3. (a) 20.46   (b) 49.59   (c) 127456.23

4. 0.375

Chapter 5

Progress questions *(page 70)*

1. 1 $a^{b/c}$ 2 =   2. To find the square root of a number.

3. Usually by using the M+ button.

4. To recall the answer from the previous calculation.

5. 5 x^y 4 =

Chapter 6

Practice questions *(page 76)*

(a) $x = 50°$ (b) $x = 95°$ (c) $x = 110°$

(d) $a = 60°$ $b = 120°$ $c = 60°$

(e) $e = 55°$ $f = 90°$ $g = 55°$ $h = 35°$

(f) $i = 70°$ $j = 80°$ $k = 70°$ $l = 30°$

(g) $q = 90°$ $m = 60°$ $n = 90°$ $o = 30°$

Practice questions *(page 79)*

(a) 35°   (b) 60°   (c) 50°

(d) 60°   (e) 70°

Progress questions *(page 88/89)*

1. Acute

2. (a) $x = 60°$ $y = 120°$ $z = 60°$

    (b) $a = 45°$ $b = 135°$ $c = 45°$ $d = 135°$

        $e = 45°$ $f = 135°$ $g = 45°$

Chapter 7

Practice questions *(page 94)*

1. 268.7cm^2 2. 192m^2 3. 485m^2 4. (a) 4m (b) 18m^2

Practice questions *(page 97)*

1. 113.09cm^2 2. 31.4cm   3. 37.7cm

4. 78.54m^2 5. 452.39m^2

Progress questions *(page 104)*

1. (a) 452.39cm^2 (b) 201.06cm^2

2. 32m^2

3. 9047.79cm^3

Chapter 8

Practice questions *(page 112)*

1. (a) mean = 7, median = 7, mode = 4

    (b) mean = 7, median = 9, mode = 3

2. Modal temp = 4

3. Both have the same mean score; therefore it could be argued that both have similar ability.

4. (a) Mean = £43 940. Median = £22 200. There is no mode.

   (b) The mean is unfair. George's salary distorts the mean.

5. Mean = 5.6, median = mode = 6.

Progress questions *(pages 127/128)*

1. (a) *See charts on page 129*

   (b) 42 cars

2. (b) It shows a positive correlation.

Chapter 10

Practice questions *(pages 164-66)*

A.

1. $x = 12$ 2. $y = 5$ 3. $x = 5$ 4. $t = 7$

5. $x = 9$ 6. $x = 3$ 7. $x = 7$ 8. $x = 5$

9. $x = 5$ 10. $x = \frac{4}{7}$ 11. $x = \frac{3}{4}$ 12. $x = 1$

13. $x = 42$ 14. $x = 35$

B.

1. $x = c + b - m$ 2. $x = \frac{v+j}{k}$ 3. $x = \frac{y}{l} - m$

4. $x = b - \frac{h+f}{n}$ 5. $x = (a+b)(m+n)$

6. $x = \frac{y}{d} - a$ 7. $\frac{y}{t} = x$ 8. $x = 2y$

9. $x = f + \frac{d}{f}$ 10. $x = \frac{az}{m} - d$

C. The mistake is in the second line. The writer should have subtracted $2x$ from both sides and not added $2x$ to both sides.

D. The mistake is in the fourth line of working; when both sides are multiplied by x, that means all the terms should be multiplied by x. So the line should read:

$$\frac{4}{x} \times x = xc - xd(1 + f) - x$$

Glossary

Acute: any angle less than $90°$ is classified as an acute angle.

Algebra: the branch of mathematics that deals with the general case. For example, using Pythagoras, $5^2 = 3^2 + 4^2$ is a specific case whereas $c^2 = a^2 + b^2$ is a general case. Algebra involves the use of letters to represent variables and is an extremely powerful tool in problem solving.

Angle: the measure of space between two intersecting lines.

Arc: part of the circumference of a circle.

Area: the measure of two-dimensional space. This is usually measur-ed in square units.

Average: *see* mean, median, mode

Bar chart: a chart that is used to display the distribution of a set of data. The height of the bar is proportional to the frequency.

Bearing: a measure of location on a compass, usually given as a three figure bearing rotating clockwise from North, where North has the bearing $000°$.

Chord: A straight line across a circle from one part of the circumference to another part of the circumference, but NOT passing through the centre point.

Circumference: The boundary or perimeter of a circle.

Correlation: this is a connection between two variables. It can be a positive correlation; in other words, as one of the variables increases, the other increases in the same proportion. Alternatively it could be a negative correlation, where as one variable increases, the other decreases in the same proportion. An example of a positive correlation is lung cancer and smoking, the incidence of lung cancer is high amongst smokers, this indicates they are connected.

Compound interest: the type of interest that is paid by most banks, where the interest is added to the principle invested and then in subsequent years, interest is earned by the original interest.

Cylinder: a three-dimensional shape which is classed as a prism. The volume of the cylinder is found by calculating the area of cross section, in this case a circle, and then multiplying this by the length.

Denominator: bottom part of a fraction.

Diameter: the distance from one point on the circumference of a circle, drawn as a straight line, to another point on the circumference, passing through the centre point of the circle.

Dimensions: a dimension is a length. Formulae can be analysed to

197

determine what type of formulae they may be, for example, $A = lb$, where l and b are both lengths, could be an area. This is because a formula for area must contain two lengths. Similarly, a formula for volume must contain three lengths.

Distribution: a set of data is called a distribution.

Equation: this is a mathematical statement in algebra, where two sides of the statement are equal. The objective is to determine the value of the unknown by manipulating the equation.

Equation, linear: this is an equation where the highest power of any of the terms is one.

Equation, quadratic: this is an equation where the highest power of any of the terms is two.

Equilateral triangle: a triangle with all three sides and all three angles of equal length. Therefore, since there are 180° in a triangle, each angle of an equilateral triangle must equal 60°.

Estimate: the process of comparing the size of a property of an object, with a known quantity.

Euro: the name of the currency introduced to Europe in 1999. This currency will form the common currency of a number of countries.

Factor: a number that divides exactly into another number with no remainder. For example, the factors of 6 are 1, 2, 3, 6 because all of these numbers divide exactly into 6, with no remainder. In algebra, a factor can be a term, as well as a number, for example, x is a factor of x^2 because $x \times x = x^2$.

Formula: a statement, usually written in algebra, which is the result of previously established accepted work. This formula is accepted as being correct and can be used in subsequent work. For example, to find the circumference of a circle, one available formula is $C = \pi d$.

Frequency polygon: a way of displaying grouped data, where the mid-values of the class intervals are joined by lines.

Function: a function is a rule which applies to one set of quantities and how they relate to another set.

Generalising: a process in mathematical thinking where a general rule, usually expressed in algebra, is determined.

Gradient: the measurement of the steepness of the slope of a line. It is the ratio of the vertical to the horizontal distance.

Graphical calculator: an electronic calculator on which it is possible to draw graphs. This type of machine is becoming common in many schools and colleges.

Graph: a mechanism for displaying data distributions to create a visual understanding of the nature of the distribution.

Highest common factor: the highest factor that will divide exactly into two or more numbers. For example, the highest common factor of 8 and 12 is 4, because 4 is the highest number that will divide exactly into 8 and 12.

Hypotenuse: the longest side of a right-angled triangle.

Imperial units: traditional units of measure used historically in Britain. Feet and inches, gallons, pints and miles continue in use, but many measures such as rods, poles, perches, furlongs and chains are no longer used in everyday life.

Improper fraction: a fraction where the top of the fraction (the numerator) is of a higher value than the lower part of the fraction, (the denominator).

Independent events: events in probability that are not dependent on previous experience. For example, when a coin is spun through the air, the chance of it landing as a head is independent of the number of spins.

Interquartile range: the spread of the middle 50% of a distribution.

Isosceles triangle: a triangle with two equal sides and two equal angles and where the equal angles are opposite the equal sides.

Line of symmetry: a line that bisects a shape so that both parts of the shape are reflections of each other.

Lowest common multiple: the lowest number that two or more numbers will go into. For example, the lowest common multiple of 5 and 10 is 10, because 10 is the lowest number that both 5 and 10 will divide into exactly.

Mean: the arithmetic average, determined by adding all of the items of data and then dividing by the number of items.

Median: a second type of average, it is the middle value in a set of data, when they are arranged in order of size, from highest to lowest or from lowest to highest.

Metric: the system of measures commonly used in Europe and the most commonly used system in science.

Mode: this is the value that occurs the most frequently. It is the average that is used in the clothes industry, when one talks of the average size man or the average size woman.

Negative number: a number with a value less than one.

Net: the pattern made up by a three-dimensional shape, when it is cut into its 'construction template' and laid flat.

Number, cube: the sequence of cube numbers is 1, 8, 27, 64, 125 etc. and is made up from $1 \times 1 \times 1$, then $2 \times 2 \times 2$, then $3 \times 3 \times 3$ etc.

Number, prime: a prime number is defined as a number that has two factors only, and those factors are different. The factors are the number itself and one.

Number, square: the sequence of square numbers is 1, 4, 9, 16, 25, 36, 49, 64 etc and is formed by 1 × 1, then 2 × 2, then 3 × 3 etc.

Numerator: top part of a fraction.

Obtuse: any angle greater than 90° but less than 180° is called obtuse.

Parallel lines: these are classically defined as lines that are equidistant along the whole of their length; in other words, they remain a constant distance apart and never meet. This is true, although there are certain circumstances that are studied in higher mathematics where this definition can be challenged, but it will suffice for our purposes.

Percentage: a fraction over 100.

Perimeter: the total distance around the boundary of a shape.

Perpendicular lines: these are lines that meet at right angles.

Pie chart: a chart in the shape of a circle, or pie, where the size of the sector is indicative of the frequency.

Polygon: a many-sided shape.

Powers (also called indices): the power to which a number is raised.

Prism: a solid of uniform cross-section.

Probability: the mathematical study of the chance of an event occurring.

Pythagoras' theorem: this is a statement that indicates the relationship that exists between the square of the hypotenuse of a right-angled triangle and the other two sides of the triangle. Classically the theorem is stated as: 'The square on the hypotenuse of a right-angled triangle is equal to the sum of the squares on the other two sides.'

Quadratic equation: *see* equation.

Quadrilateral: a plane, four-sided shape.

Radius: the distance from the centre of a circle to any point on the circumference. The radius of a circle is therefore half of the diameter of the same circle.

Ratio: the relationship between one quantity and another, expressed as the number of times that one quantity can be divided by the other.

Reflex: any angle greater than 180° belongs to this group of angles.

Right angle: an angle of 90°.

Scalene triangle: a triangle where all of the sides and the angles are of different sizes.

Sequence: a set of numbers having a common property which, when deduced, allows the reader to predict further numbers in the sequence.

Trapezium: a quadrilateral — the shape that is left when the top of a triangle is cut off.

Units: *see* Imperial, and *see* Metric.

V.A.T.: Value Added Tax.

Index

Also by Dr Graham Lawler

Understanding the Numbers

The first steps in managing your money

2nd EDITION

Financial Literacy for troubled times

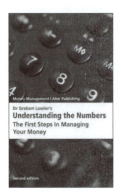

- **<u>Does for financial literacy what Lynn Truss did for grammar.</u>**

- A growing market of mathematically illiterate adults – in need of bite - sized explanation

- Dr Graham Lawler is an articulate broadcaster with regular slots on radio

Who else wants to make money?

Understanding the Numbers takes a practical rather than a theoretical approach, based on the everyday financial situations we all have to face, the author has followed on from his highly successful *Understanding Maths* book and introduces the reader to basic financial literacy:

Author

Dr Graham Lawler

Price   £11.99

Format

Paperback,
215 x 135mm, 160pp

ISBN

978-1-84285-125-8

This book includes details on how to:

- deal with decimals and fractions

- understand cashflow and cash analysis

- re-engineer personal finances

- solve the four types of % problems

- analyse and interpret data

- solve simple linear equations

<u>About the Author</u>

Dr Graham Lawler (aka 'Mr Educator') is the author of Understanding Maths, *BBC Bitesize Maths,* and *Understanding Algebra* and broadcasts regularly on radio as 'Mr Educator'. He has acted as a consultant on mathematical education for both the BBC and Channel 4, and is now a company director.

This book is ideal for:

- Adults wishing to use maths to improve their financial standing and for courses, jobs or just to recover lost ground
- First-year MBA (an essential read) and GMAT students
- Parents who wish to support their children's education
- Business and FE colleges and school libraries
- Public libraries.

SUBJECT AREA
Mathematics/Education/ Basic Finance

Related Titles
Understanding Maths 1-84285-086-5
Algebra 1-84285-068-7

Algebra
Basic algebra explained

Algebra

Basic algebra explained

For post - 16 support, BTEC Students, adult education, engineers, technicians, nurses and GMAT Students.

Information Points

- A growing market of mathematically under-performing adults – ready to snap up this book.
- Crossover markets – the MBA and business market
- Ideal for primary student teachers and student nurses

In Focus – A Studymates Series

Author

Dr Graham Lawler

Price

£10.99

Format

Paperback,

215 x 135mm,

ISBN

978-1-84285-068-8

Who else wants to understand and use basic algebra?

This book covers the essential information needed to understand the basic concepts in algebra and how to use algebraic techniques to solve problems.
Are you enrolled on a course where simple algebraic concepts are required? Or are you a parent struggling to keep up with your child´s homework?

Algebra: basic algebra explained takes a practical rather than a theoretical approach, and explains the basic skills that all algebra users need to know.

The book explains how to:

- solve linear equations
- factorise quadratic equations
- change the subject of a formula
- solve equations with fractions
- solve simultaneous linear equations
- understand straight line graphs

Market

- Mathematics or business students, at A-Level or at an FE Collage
- First-year undergraduates
- MBA and GMAT students
- Parents
- Business and FE colleges *
- Public libraries.

RELATED TITLES
Understanding Maths, Basic Mathematics explained Back to the black: How to get out of debt and stay out of dept

Money Management: An Aber Series

Back to The Black

How to get out of debit and stay out of debt

Who else wants to clear their debts?

Back to the black takes a practical rather than a theoretical approach, based on the everyday financial situations we all have to face. The author has followed on from his highly successful *Understanding the Numbers* book, the first in this series, and introduces the reader to basic financial literacy:

Author

Dr Graham Lawler

Price £11.99

Format

Paperback,
215 x 135mm, 160pp

ISBN

978-1-84285-141-8

This book includes details on how to:

- quantify your debt
- understand cashflow and cash analysis
- re-engineer personal finances
- pay yourself first
- analyse and interpret payment schedules
- solve your own financial problems

This book is ideal for:

- Adults wishing to use maths to improve their financial standing and clear their debts
- Students of basic finance courses
- Parents who wish to support their children's education
- Business and FE colleges
- Public libraries.

SUBJECT AREA
Mathematics/Education/ Basic Finance

Related Titles
Understanding the Numbers 978-1-084285-125-8
Understanding Maths 978-1-84285-086-2
Algebra 978-1-84285-068-8

Keep up to Date with Dr Lawler via his website www.graham-lawler.com

Dr Lawler also created the **British Debt to Wealth System,** *Back to the Black* so you can keep up to date with his work via his now well-known *Letter From Abergele.* You can also join his mailing list from this site. This mailing list covers *Studymates, Aber* and *Mr Educator* Books and Media and will NEVER be sold on to anyone else. Dr Lawler has instructed us to ensure that this is always the case.

You are also welcome to visit us at www.aber-publishing.co.uk or our sister site at www.studymates.co.uk or for USA readers, www.studymates-usa.com